D0758914

Ecology and Management of North American Savannas

Ecology and Management of North American Savannas

Guy R. McPherson

The University of Arizona Press

Tucson

The University of Arizona Press

⊗ This book is printed on acid-free, archival-quality paper
Manufactured in the United States of America
First printing

Library of Congress Cataloging-in-Publication Data
McPherson, Guy R. (Guy Randall), 1960-
Ecology and management of North American savannas / Guy R. McPherson.
p. cm.
Includes bibliographical references and index.
ISBN 0-8165-1624-3 (cloth : acid-free paper)
1. Savanna ecology—North America. 2. Range management—North America. 3. Savannas—North America. I. Title.
QH102.M36 1997
577.4'8'094—dc21 97-4813
CIP

British Cataloguing-in-Publication Data
A catalogue record for this book is available from the British Library.

To Sheila with love,

and to my students with affection and gratitude

Contents

Illustrations

Preface

The writing of this book was motivated by three primary factors. First, there is a substantial and widely scattered literature on North American savannas that has not been summarized, synthesized, or integrated. Second, I would like to encourage constructive debate and relevant research on these systems because I believe that they are fascinating and important. Third (and the most selfish), I wanted to spend my sabbatical leave on an intellectually stimulating project that would force me to go beyond simply "catching up" with individual research projects. There is no question that the project satisfied the latter of these goals; individual readers must decide if this book accomplishes either of the first two.

North American savannas represent a diverse and disparate group of ecosystems. A central tenet of this book is that despite the dissimilarities among savannas, the structure and function of savannas are sufficiently similar to merit synthesis. Thus, the primary intent of the book is to provide an overview of the patterns and processes shared by these ecosystems. Chapter 1 provides some definitions and boundaries and describes each of the major savannas covered in the book. The discussion proceeds at a relatively fine scale of resolution in chapter 2, then expands to encompass larger spatial and temporal scales in chapters 3 through 5. The book provides a consideration of ecological applications in chapter 6 and concludes with a chapter on needed research.

The book is targeted at ecologists, natural resource managers, and sophisticated laypeople who have an interest in savannas. It may be particularly useful for students and researchers who are beginning to study savannas or who want a general overview.

Because this book represents the only synthesis of the expansive literature on North American savannas, it also may provide a useful comparative context for researchers on other continents. Although it was not written specifically as a course textbook, the book may prove useful as a supplemental text in courses dealing with rangeland ecology, plant community ecology, biogeography, or natural resource management. Upper-division undergraduate students and graduate students in ecology and natural resource programs should have sufficient background to read and understand the concepts discussed herein.

Acknowledgments

My wife, Sheila Merrigan, has provided guidance, support, and assistance in ways too numerous to list. This book is a joint venture; it would not have been possible without her.

My graduate students motivate me with their interest, dedication, and enthusiasm. They are the primary reason I love my work. I owe particular thanks to Kathryn Haworth, Charles Nyandiga, Sam Drake, Margi Brooks, Svenje Mehlert, Andy Hubbard, Jose Villanueva-Díaz, Laurie Abbott, Paulette Ford, Jake Weltzin, and Heather Germaine.

Numerous other students have worked in the lab and contributed to this effort through their dedication and attention to detail. Alicia Nicholas, Karin Rojahn, Kim Suedkamp, Anastasia Olander, Rob Matson, and Pete Bergman deserve special thanks for their efforts.

I appreciate the support and good humor of my colleagues at the University of Arizona. I feel fortunate that many of these colleagues have become good friends in the eight years I have been in Tucson. I am particularly indebted to Lisa Graumlich, Bill Mannan, Bill Matter, and Mitch McClaran.

Most of the manuscript was prepared while I was on a sabbatical leave at the University of California. Jim Bartolome hosted my visit and contributed greatly to my productivity while in Berkeley. Financial support for the sabbatical was provided by the University of Arizona, University of California, and USDA Forest Service Rocky Mountain Experiment Station Borderlands Research Project.

In addition to sharing ideas, several people unselfishly provided photographs and unpublished data and manuscripts during the preparation of this manuscript. Particularly worthy of mention are Steve Archer, Bill Van Auken, Jim Bartolome, Barbara Allen-Diaz, Lynn Huntsinger, Conrad Bahre, Fred Smeins, Heather Germaine, Bill Platt, and Lindsay Boring.

Constructive reviews of parts or all of the manuscript were provided by Steve Archer, David Wester, Jake Weltzin, Heather Germaine, Laurie Abbott, Keirith Snyder, Dave Williams, Keith Blair, Lindsay Boring, Bill Platt, and Sheila Merrigan. Their efforts greatly improved the manuscript. Particularly heroic were the efforts of Jake Weltzin, who read several drafts of the manuscript and provided insightful, critical reviews. His mini-essays and sketches in the margins were always interesting and occasionally humorous: both attributes provided welcome relief from the onerous task of revising the manuscript.

Amy Chapman Smith at the University of Arizona Press solicited the manuscript and kept me on track during the early stages of this project. Chris Szuter picked up where Amy left off. I appreciate the skill and patience of both editors.

Few of the ideas in this book are uniquely mine. I have borrowed them from colleagues, many of whom are mentioned in the preceding paragraphs. I thank them for their insight and ask their forgiveness for losing track of who had the ideas first. Errors of fact or interpretation remain my own.

Finally, my family has provided moral, emotional, and logistical support not only during the preparation of this manuscript, but throughout my career. They have shaped my values and personality, for which I am indebted. Each of them is a friend as well as a family member: my parents, Jim and Eddy McPherson, and siblings, Jim McPherson and Carol Wallace.

Ecology and Management of North American Savannas

1

Importance and Extent

What is there about the Savannah?
T. L. Stokes, The Savannah

North American savannas, defined as ecosystems with a continuous grass layer and scattered trees or shrubs, are extensive and diverse. They range from xeric pinyon-juniper communities in the Great Basin to subtropical longleaf pine habitats along the Atlantic coastal plain. Dominant woody plants include species of important economic and cultural value as well as species commonly regarded as weeds. Despite the expansive areal extent and importance of savannas, ecological relationships and management practices for savannas are incompletely defined compared to those of other physiognomic types in North America (e.g., forest, desert, grassland). Thus, the primary goals of this book are to synthesize a rapidly growing literature on savanna ecology and management, discuss ecological patterns and processes, and develop a framework for implementing management practices in North American savannas.

Savannas are important to humans for a variety of cultural and commercial reasons. Savannas occupy nearly a third of the world's land surface (Werner et al. 1990), including over 50 million hectares of North America. Savannas represent an important source of natural resources, including nonconsumptive (e.g., many types of recreation) and consumptive (e.g., wood products, forage for livestock) uses. Considerable evidence indicates that hominids evolved in savannas (Harris 1980, Schule 1991). Further, savannas represent a preferred vegetation type around dwellings and settlements: humans typically create savannas by removing trees from forests and adding trees to grasslands and deserts (Ferris-Kaan 1995).

The proportion of woody plant cover in savanna varies from

less than 1% to about 30%. Specification of a precise amount of woody plant cover is not imperative; however, recognition of two distinct layers of vegetation—a woody overstory and a graminoid understory—is central to the savanna concept. This interpretation of savanna, which is based on vegetation physiognomy, is well accepted in plant ecology and vegetation science (e.g., Eiten 1986, 1992) and will be used throughout this book. Communities dominated by scattered woody plants but having an understory of low-growing shrubs are not considered savannas. For example, northern pinyon-juniper woodlands are characterized by understories dominated by sagebrush and grasses and thus are not classified as savannas. However, many southwestern pinyon-juniper types fit the definition of savanna.

The contemporary definition of savanna differs from the original use of the term with respect to woody plant cover: the original definition—a treeless plain—was synonymous with what we now consider as prairie. The term "savanna" was acquired by the Spaniards from Taino, the language of an extinct group of Arawak Indians from the Greater Antilles and Bahamas (Frost et al. 1986). The term "savanna" was initially used in North America to describe forest openings or prairie (e.g., Bartram 1791). The term "barrens," which is synonymous with "savanna" in contemporary usage, has been used to describe forest openings in the eastern United States (Heikens and Robertson 1994, 1995, Hutchison 1994). The definition of the terms "savanna" and "barrens" gradually diverged from the concept of "treeless plain," but "savanna" was rarely used by North American ecologists until 1957, when E. J. Dyksterhuis published a seminal paper on use of the term (Dyksterhuis 1957).

Despite their importance, savannas (particularly temperate savannas) are generally under-studied compared to other physiognomic types. Tropical savannas in Africa, South America, southeastern Asia, and northern Australia have begun to receive more attention from the scientific community within the last two decades, presumably because of widespread concern about tropical biodiversity and conservation (e.g., Harris 1980, Tothill and Mott 1985, Walker 1987, Young and Solbrig 1993, Solbrig et al. 1996). In contrast to their tropical counterparts, temperate savannas remain under-studied, and no synthesis of North American savanna literature has previously been attempted (but see Anderson et al.

1998). This scientific neglect may stem from the absence of a professional discipline associated with savannas (cf. forestry's focus on forests and range management's focus on grasslands), limited understanding of the role and importance of savannas in temperate regions, or inconsistent definitions or interpretations of the term "savanna." The latter factor is exemplified by definitions of savanna adopted by tropical researchers on the basis of climate (i.e., possessing distinct wet and dry seasons) or geography (i.e., occurring only in tropical latitudes) (e.g., Sarmiento 1984, Tothill and Mott 1985, Walker 1987, Young and Solbrig 1993, Solbrig 1996). Language is a powerful tool for advancing or impeding scientific activities: defining savannas on the basis of tropical climate or equatorial geography essentially precludes savannas from temperate regions, leaving the impression that savannas are absent from North America.

Savannas and Natural Resources

Large-scale benefits derived from savannas are seldom recognized or appreciated. However, if viewed at a coarse level of resolution, savannas are natural ecosystems that provide a variety of environmental "services." The "free services" provided by savannas and other natural ecosystems led Eugene Odum (1989) to call them the "life-support system" for urban areas: they provide clean air and water, open space, and biodiversity. In addition, savannas form an important component in the global carbon cycle: woody plants and deep-rooted grasses in savannas may serve as significant long-term global storage compartments for carbon, helping mediate anthropogenically induced increases in atmospheric CO_2 concentrations (McPherson et al. 1993, Nepstad et al. 1994). Further, given the globally extensive areas of grassland and savanna undergoing significant increases in woody plant biomass, increased carbon storage in stems, roots, and the soil beneath woody plants could be a significant component of the global carbon cycle (McPherson et al. 1993).

Resources derived from savannas at the regional or local level are numerous, varied, and usually widely recognized and appreciated by local peoples. Native peoples have used savanna species as sources of food, fuel, medicine, and tools for generations. Because technological advances have generated substitutes for most of these products, the next section focuses on dominant contemporary resources.

Benefits of Savanna Physiognomy

The greatest use of savannas probably results from their broad-scale physiognomy rather than from attributes of the individual species that are present in any particular savanna. For example, their high aesthetic value is appreciated by hikers, picnickers, mountain-bikers, off-road vehicle users, and campers. In addition, savanna physiognomy provides habitat for many highly visible animal species, which attracts bird-watchers and hunters. Finally, most savannas provide perennial surface water, which enhances aesthetics and biodiversity and provides fishing opportunities and water for terrestrial animals.

Resources Derived from Woody Plants

Wood products derived from North American savannas include fuelwood, structural lumber, pulpwood, charcoal, barbecue chips, and Christmas trees. Virtually all savannas near urban areas are subject to a high demand for fuelwood regardless of which woody species are present. Oak and mesquite are particularly noteworthy sources of fuelwood and charcoal because of their high heat content (about 30% more energy per unit volume than average conifers). Pine savannas in the southeastern and western United States are significant sources of structural lumber and pulpwood. Additional pulpwood is derived from pinyon-juniper savannas in the Great Basin region. In addition, savanna species are locally important sources of fenceposts, furniture, specialty products, and Christmas trees.

In addition to being a significant source of wood products, savanna trees also provide raw material for other products. For example, pine needles (i.e., leaves) are baled and sold as mulch for planting-beds and gardens in the southeastern United States.

Leaves from long-needled pines (e.g., longleaf pine, ponderosa pine) are woven into decorative baskets.

Finally, the seeds of several savanna trees provide food for humans. Pinyon seeds are marketed as "pine nuts" in the western United States and northern Mexico. Acorns (particularly from Emory oak) are eaten raw or roasted for snack food, and acorns and mesquite seeds are ground into flour for cooking in northern Mexico and, to a lesser extent, in the southwestern United States. Historical uses of oak and mesquite by Native Americans were far more varied than current uses (Nabhan 1985).

Resources Derived from Understory Plants

Production of understory plants is constrained by temperature and soil moisture. For example, understory production in annual grass-dominated Californian oak savannas is restricted to the winter, when moderate temperatures are accompanied by high soil moisture (ca. December–April); the hot, dry mediterranean climate limits herbaceous production during the summer and autumn. In contrast, understory production in perennial grass-dominated southwestern oak savannas occurs during the summer, coincident with the summer monsoon (ca. July–September); in these savannas, herbaceous production is constrained by cold temperatures during the winter.

Livestock grazing is the dominant land use of North American savannas. With the exception of several parks and a few research natural areas, virtually all North American savannas are grazed by livestock. The goal of livestock grazing is to convert herbaceous biomass to a form suitable for human consumption; in addition, livestock grazing is at the core of the long heritage of the ranching industry that has been important in the settling of these savannas. Relative to other uses of savannas, livestock grazing is an extensive management enterprise: labor and capital inputs per unit of land area are low. Furthermore, several other uses (discussed throughout this chapter) are compatible with livestock grazing and occur simultaneously.

Additional resources derived from understory plants are commercially important at local scales. Examples include beargrass (used for making baskets and brooms), yucca (baskets, cattle fodder), agaves (mescal, baskets, food), numerous species transplanted for landscaping, and chiltepines and mushrooms (food).

Major Savannas

North American savannas occur throughout the western United States, in northern Mexico, and in the coastal region of the southeastern United States. They are also represented by fragments of midwestern oak savannas in the Great Plains; these fragments are scattered from near the U.S.-Canada border to central Texas. Thus, savannas are found from coast to coast throughout much of the southern half of North America.

Savannas are relatively contiguous on the continent (fig. 1.1). In fact, all major North American savannas border mesquite savanna in at least one part of their range. The southern extent of midwestern oak savannas is bordered by mesquite savannas in Oklahoma and Texas. Patches of pinyon-juniper savannas are found on coarse-textured soils throughout mesquite savannas, and pinyon-juniper savannas are well represented along the northern boundary of mesquite savannas. Ponderosa pine savannas are intermingled with mesquite savannas in the foothills of the southern Rocky Mountains and the northern Sierra Madre. The southern limit of Californian oak savannas is adjacent to the northwestern limit of mesquite savannas. Finally, mesquite savannas overlap with the natural range of longleaf pine savannas in southern Texas, although longleaf pine savannas have been largely extirpated from the region.

In contrast to our understanding of the contemporary distribution of savannas, the distribution of savannas in the geologic past is not well understood—primarily because the fossil record used to study changes in the distribution of species does not indicate changes in vegetation physiognomy. Nonetheless, interpretation of fossil records suggests that savannas were a dominant physiognomic type in North America by the late Miocene, between 5 million and 8 million years ago (P. S. Martin 1975, Axelrod 1979, Watts 1980, Van Devender 1995). Post-Cretaceous development of regional floras suggests that contemporary floras of the Sierra Madre, Great Basin, and California have a common ancestor, the Madro-Tertiary geoflora, whereas longleaf pine savannas apparently were derived from the Arcto-Tertiary geoflora (Vankat 1979).

The dominant woody taxa in North American savannas have been present for at least 10 million years, and some have dominated regional floras for over 70 million years (Axelrod 1937, 1950, 1958, 1978, Raven and Axelrod 1978, Delcourt and Delcourt 1987). However, distribution patterns of particular species have varied substantially through geologic time, presumably because of fluctuating climates. This phenomenon is exemplified by a dramatic increase in the areal extent of southeastern pines within the last 4,000 years (Watts 1969, Delcourt and Delcourt 1987). Modern cool-season grasses (with the C_3 photosynthetic pathway) became widely distributed and abundant in North America about 24 million years ago, and modern warm-season grasses (with the C_4 photosynthetic pathway) followed, between 5 million and 8 million years ago (Axelrod 1979, Ehleringer et al. 1991, Van Devender 1995).

Pinyon-Juniper Savannas

Distribution. The pinyon-juniper savanna is a subset of the more extensive pinyon-juniper woodland, which may be dominated by any of several species of pinyon or juniper. Pinyon-juniper woodlands occupy about 24 million hectares in the western United States and 5 million hectares in the Sierra Madre of Mexico. Most of these woodlands are not considered savannas, either because woody plant cover exceeds 40% or because grass cover is discontinuous; thus, one-third of the woodland area, or about 10 million hectares, is estimated to be savanna. Dominance by pinyon pine is rare, and pinyon is often absent from savannas. Nonetheless, historical precedence will be followed in labeling these savannas and associated woodlands "pinyon-juniper."

Pinyon-juniper savannas represent the warm, dry limits of coniferous vegetation and are found primarily on foothills, low mountains, mesas, plateaus, and rolling plains (fig. 1.2). Elevation is typically between 1,400 and 2,000 m in the southern Rocky

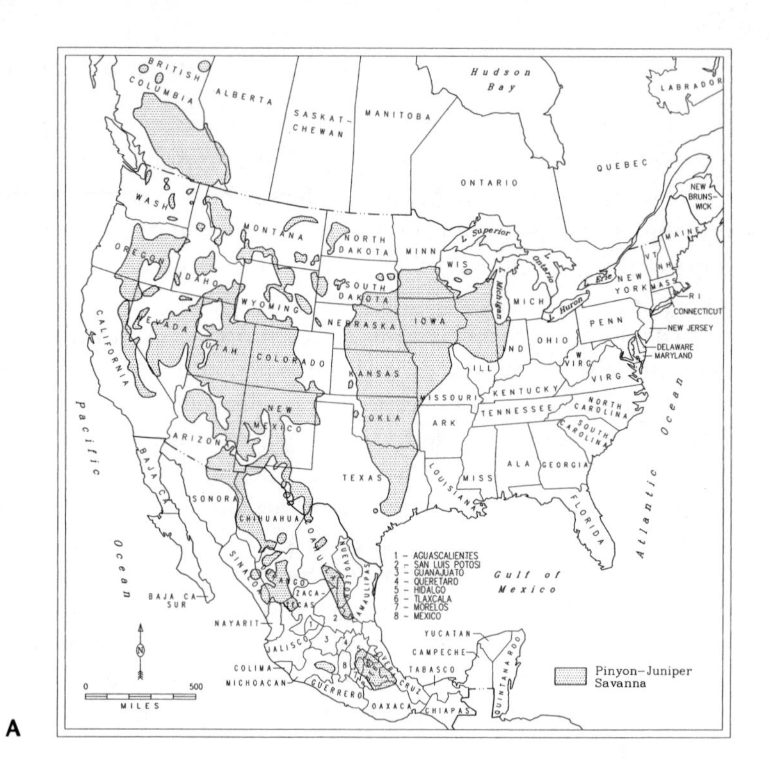

A

BRITISH COLUMBIA
ALBERTA
SASKAT-CHEWAN
MANITOBA
Hudson Bay
LABRADOR
QUEBEC
ONTARIO
NEW BRUNS-WICK
WASH
MONTANA
NORTH DAKOTA
MINN
L. Superior
MAINE
OREGON
IDAHO
SOUTH DAKOTA
WIS
L. Michigan
L. Ontario
NEW YORK
MASS
RI
CONNECTICUT
WYOMING
IOWA
MICH
Erie
Huron
PENN
NEW JERSEY
NEVADA
CALIFORNIA
UTAH
COLORADO
NEBRASKA
IND
OHIO
ILL
W VIRG
DELAWARE
MARYLAND
KANSAS
MISSOURI
KENTUCKY
VIRG
Pacific
Ocean
ARIZONA
NEW MEXICO
OKLA
ARK
TENNESSEE
NORTH CAROLINA
SOUTH CAROLINA
Atlantic
Ocean
BAJA CA
TEXAS
MISS
ALA
GEORGIA
LOUISIANA
FLORIDA
SONORA
CHIHUAHUA
COAHUILA
NUEVO LEON
TAMAULIPAS
SINALOA
DURANGO
ZACA-TECAS
BAJA CA-SUR
1 - AGUASCALIENTES
2 - SAN LUIS POTOSI
3 - GUANAJUATO
4 - QUERETARO
5 - HIDALGO
6 - TLAXCALA
7 - MORELOS
8 - MEXICO
Gulf of Mexico
NAYARIT
JALISCO
YUCATAN
CAMPECHE
QUINTANA ROO
TABASCO
VERACRUZ
COLIMA
MICHOACAN
GUERRERO
OAXACA
CHIAPAS
N
0
500
MILES
Californian Oak Savanna
Southwestern Oak Savanna
Midwestern Oak Savanna

C

Hudson Bay
LABRADOR
BRITISH COLUMBIA
ALBERTA
SASKATCHEWAN
MANITOBA
QUEBEC
ONTARIO
NEW BRUNSWICK
WASH
OREGON
MONTANA
NORTH DAKOTA
SOUTH DAKOTA
MINN
WIS
MICH
L. Superior
L. Michigan
L. Huron
L. Erie
L. Ontario
NEW YORK
MAINE
VT
NH
MASS
R I
CONNECTICUT
NEW JERSEY
DELAWARE
MARYLAND
IDAHO
WYOMING
NEVADA
UTAH
COLORADO
NEBRASKA
IOWA
ILL
IND
OHIO
PENN
W VIRG
VIRG
KANSAS
MISSOURI
KENTUCKY
TENNESSEE
NORTH CAROLINA
SOUTH CAROLINA
CALIFORNIA
ARIZONA
NEW MEXICO
OKLA
ARK
ALA
MISS
GEORGIA
LOUISIANA
FLORIDA
TEXAS
Pacific Ocean
Atlantic Ocean
Gulf of Mexico
BAJA CA
SONORA
CHIHUAHUA
COAHUILA
NUEVO LEON
TAMAULIPAS
SINALOA
DURANGO
ZACATECAS
BAJA CA- SUR
NAYARIT
JALISCO
COLIMA
MICHOACAN
GUERRERO
OAXACA
CHIAPAS
VERA CRUZ
TABASCO
CAMPECHE
YUCATAN
QUINTANA ROO
1 – AGUASCALIENTES
2 – SAN LUIS POTOSI
3 – GUANAJUATO
4 – QUERETARO
5 – HIDALGO
6 – TLAXCALA
7 – MORELOS
8 – MEXICO
N
0
500
MILES
Longleaf Pine Savanna
Ponderosa Pine Savanna

B

Hudson Bay
LABRADOR
BRITISH COLUMBIA
ALBERTA
SASKATCHEWAN
MANITOBA
QUEBEC
ONTARIO
NEW BRUNSWICK
WASH
OREGON
MONTANA
NORTH DAKOTA
SOUTH DAKOTA
MINN
WIS
MICH
L. Superior
L. Michigan
L. Huron
L. Erie
L. Ontario
NEW YORK
MAINE
VT
NH
MASS
R I
CONNECTICUT
NEW JERSEY
DELAWARE
MARYLAND
IDAHO
WYOMING
NEVADA
UTAH
COLORADO
NEBRASKA
IOWA
ILL
IND
OHIO
PENN
W VIRG
VIRG
KANSAS
MISSOURI
KENTUCKY
TENNESSEE
NORTH CAROLINA
SOUTH CAROLINA
CALIFORNIA
ARIZONA
NEW MEXICO
OKLA
ARK
ALA
MISS
GEORGIA
LOUISIANA
FLORIDA
TEXAS
Pacific Ocean
Atlantic Ocean
Gulf of Mexico
BAJA CA
SONORA
CHIHUAHUA
COAHUILA
NUEVO LEON
TAMAULIPAS
SINALOA
DURANGO
ZACATECAS
BAJA CA- SUR
NAYARIT
JALISCO
COLIMA
MICHOACAN
GUERRERO
OAXACA
CHIAPAS
VERA CRUZ
TABASCO
CAMPECHE
YUCATAN
QUINTANA ROO
1 – AGUASCALIENTES
2 – SAN LUIS POTOSI
3 – GUANAJUATO
4 – QUERETARO
5 – HIDALGO
6 – TLAXCALA
7 – MORELOS
8 – MEXICO
N
0
500
MILES
Mesquite Savanna

D

Mountains and northern Sierra Madre and slightly higher in the Great Basin and northern Rocky Mountains. In these regions, savannas merge with closed-canopy woodland or forest at their mesic upper limit and with grass- or shrub-dominated communities below (Evans 1988). In the central United States, pinyon-juniper savannas occur on sites between 300 and 600 m elevation, and common adjacent communities are prairies, mesquite savannas, cultivated fields, and juniper woodlands.

Pinyon-juniper savannas are poorly developed in the northern portion of the range, where the winter-dominant precipitation regime favors the development of low-growing shrubs in the understory. In contrast, they are well represented in the southwestern United States and northern Mexico. Three million hectares of eastern redcedar savannas are scattered throughout the Great Plains, with the greatest representation in Oklahoma (Schmidt and Stubbendieck 1993). In total, pinyon-juniper savannas are represented by approximately 10 million hectares in North America.

Pinyon-juniper savannas have increased in area since Anglo settlement, and nearly all of these range expansions represent the conversion of former grasslands to savannas (West 1984, Jameson 1987, West and Van Pelt 1987). However, the density and cover of pinyon and especially juniper trees have also increased, thereby converting large areas of pinyon-juniper savanna to closed-canopy woodland (Evans 1988, Bahre 1991, Miller and Wigand 1994). Thus, the net change in area occupied by pinyon-juniper savannas is difficult to determine.

Climate, Geology, and Soils. Average annual precipitation is 250–400 mm in pinyon-juniper savannas of western North America

Figure 1.1 Savannas are widely distributed in North America: (A) pinyon-juniper savannas; (B) long-needled pine savannas dominated by ponderosa pine in western North America and by longleaf pine in the southeastern United States; (C) oak savannas represented by Californian oak savannas in the coastal states, by southwestern oak savannas in interior southwestern North America, and by midwestern oak savannas in middle of the continent; (D) mesquite savannas. (Data sources: Wright and Bailey 1982, Nuzzo 1986, Steele 1987, Johnson and Mayeux 1990, Pavlik et al. 1991, Miller and Wigand 1994.)

Figure 1.2 Pinyon-juniper savannas usually are dominated by one species of juniper. (Photograph courtesy of Fred Smeins.)

and as high as 900 mm in the Great Plains. Seasonality of precipitation varies from winter-dominant in the northern Rocky Mountains and northern Great Basin to summer-dominant in the middle and southern Sierra Madre. In the southern Rocky Mountains and northern Sierra Madre, annual precipitation is distributed bimodally, with peaks in the summer and winter. Pinyon-juniper savannas in the central United States are characterized by precipitation that is distributed relatively evenly within the year. The frost-free period averages 120 days in the Great Basin, increasing to 300 days in the Sierra Madre and southern Great Plains.

Geologic substrates and soils are poorly correlated with the occurrence of these savannas. Soils may be derived from granite, limestone, volcanic material, basalt, sandstone, and alluvium, and they vary widely with respect to texture and structure. Soil depth ranges from 10 cm to several meters in depth, and no fewer than six soil orders are represented: Alfisols (soils with distinctive horizonation and clay accumulation), Aridisols (soils that are dry more than six months each year and that contain little organic matter in the surface layer), Entisols (soils with virtually no profile development), Inceptisols (soils with weakly developed horizons), Mollisols (well-developed soils commonly associated with grasslands), and Vertisols (soils with high clay content that are characterized by considerable mixing via shrinking and swelling). Soils vary from shallow, rocky, poorly developed sands to deep, well-developed loams (Springfield 1976, West et al. 1978, Leonard et al. 1987, Evans 1988).

Vegetation Structure and Function. Nearly all pinyon-juniper savannas are dominated by only one of the many species of juniper, although the particular juniper species varies throughout the broad range of the type. Dominant junipers, and their distribution, include western juniper *(Juniperus occidentalis)* on the lee side of the Cascade and Sierra ranges; Utah juniper *(J. osteosperma),* occasionally codominant with Rocky Mountain juniper *(J. scopulorum),* in the Great Basin; Rocky Mountain juniper in the northern and central Rocky Mountains; one-seed juniper *(J. monosperma)* in the southern Rocky Mountains; alligator juniper *(J. deppeana)* in the Sierra Madre and extreme southern Rocky Mountains; Pinchot juniper *(J. pinchotii)* in western Texas; Pinchot juniper, red-berry juniper *(J. erythrocarpa),* or drooping juniper *(Juniperus flaccida)* in the trans-Pecos region of southwestern Texas, southern New Mexico, and northern Chihuahua; Ashe juniper *(J. ashei)* in central Texas; and eastern redcedar *(J. virginiana)* in the midwestern United States (West 1984, Miller and Wigand 1994). All of these species, and associated pinyons, are evergreen conifers.

Subordinate species include other junipers, pinyons, and oaks. Single-leaf and two-leaf pinyon *(Pinus monophylla, P. edulis)* are common in the western United States, and two-leaf pinyon extends to the southern Rocky Mountains, where it merges with border pinyon *(P. cembroides)* in the Sierra Madre (Evans 1988). Dominant oaks associated with southwestern oak savannas are found in pinyon-juniper woodlands of the Sierra Madre (West 1984).

Dominant understory plants in the Great Basin and northern and central Rocky Mountains are cool-season grasses, herbaceous dicots, and shrubs. Most of the common native grasses are perennial bunchgrasses. However, introduced annual grasses dominate the understory of many of these savannas: the dominant introduced species are downy brome *(Bromus tectorum)* and medusahead *(Taeniatherum asperum)*. Farther south and east, low-growing shrubs and introduced annual grasses are replaced in the understory by warm-season grasses. These grasses include several sod-forming species in addition to bunchgrasses (Springfield 1976, West et al. 1978, West 1988).

Production of woody plants is negatively correlated with production of herbaceous plants in North American savannas (see chapters 2, 3). In pinyon-juniper savannas, average annual stemwood production is 1–2.5 m^3/ha in the western portion of the range and 2–5 m^3/ha in savannas dominated by eastern redcedar (Ferguson et al. 1968, Barger and Ffolliott 1972, Conner and Green 1988, Conner et al. 1990, Born et al. 1992, Van Hooser et al. 1993). Average annual herbaceous production in western North America ranges from less than 200 kg/ha on sites with relatively shallow soils and high juniper cover to 1,500 kg/ha on more productive sites or sites with little overstory cover (Springfield 1976, West 1984). Similarly, herbaceous production is positively correlated with soil depth and negatively correlated with overstory cover in savannas dominated by eastern redcedar or Ashe juniper, where annual production varies from 500 to 4,000 kg/ha (Wink and Wright 1973, Smeins et al. 1976, Engle et al. 1987, Smith and Stubbendieck 1990).

Land Use. The Bureau of Land Management administers about three-fourths of the pinyon-juniper savannas in the western United States, and various federal and state agencies administer most of the remainder. In contrast, over 90% of the savannas in Texas and the Great Plains are privately owned. *Ejidos* (communal landholding organizations) have management responsibility for about 70% of northern Mexico, including most pinyon-juniper savannas (Whetten 1948, Felger and Wilson 1995). The remaining Mexican lands are divided among federal and state trusts and individual owners. Mexico's constitution was modified in 1992 to allow disincorporation of *ejidos,* but widespread disincorporation has not yet occurred.

Livestock grazing is the most common use of these savannas. Livestock graze virtually all savannas in Mexico and about 80% of the savannas in the United States (Clary 1975).

Fuelwood is the most important wood product derived from

pinyon-juniper savannas. The historical importance of fuelwood harvesting has been reviewed by Bahre (1991). Private individuals, who harvest trees for personal use, account for over 95% of the fuelwood harvested (McLain 1988, 1989). As such, nearly all the fuelwood is harvested within 200 km of urban areas, because it is economically infeasible to transport it longer distances (Evans 1988). Thus, rates of fuelwood harvesting are relatively low for most savannas, particularly compared to the rates that were evident during the mining era (ca. 1850–1900). For example, less than 0.2% of the total available fuelwood is removed each year from pinyon-juniper savannas (McLain 1988, 1989, Conner et al. 1990), whereas wholesale removal of trees was common during the mining era. Fuelwood remains an important source of revenue and of energy for home heating and cooking on American Indian reservations in the United States and in remote areas of northern Mexico (Rzedowski 1983, Bahre 1991, Schwab 1993).

Other dominant uses of wood products derived from pinyon-juniper savannas are "cedar" products and fenceposts. Eastern redcedar is the primary source of "cedar" for chests, boxes, and chips (eastern redcedar is not a true cedar, i.e., it is not a member of the genus *Cedrus*). Juniper wood is widely used for fenceposts because it is resistant to decay and because appropriate-size pieces are readily available. Development of a viable tree harvesting industry in pinyon-juniper woodlands has been impeded by numerous factors, including poor marketing practices, high processing costs, and naturally low production rates compared to those of other nearby forest types (Huber 1992).

Christmas trees and pinyon "nuts" (seeds) represent additional uses of woody plants in these savannas. Pinyon pines are prized for Christmas trees because their slow vertical growth and large number of lateral branches combine to produce a conical tree with a full crown. Thus, a system of grading has been developed specifically for pinyons (Springfield 1976). Commercial operators harvest between 0.5 million and 1 million kg of pinyon nuts each year from woodlands in the western United States, and over 3 million kg are harvested during some years (Evans 1988). The harvest of pinyon nuts has been viewed as a "growth" industry for over 50 years (Little 1941, Barger and Ffolliott 1972, Delco et al. 1993), although success to date has been very limited (Little 1993). Pinyons are relatively uncommon in savannas, so pinyon-based industries are largely restricted to closed-canopy woodlands.

Recreation is a common nonconsumptive use, and it is particularly encouraged in publicly owned pinyon-juniper savannas. Hunting has long been a dominant recreational use, but people also use savannas for hiking, backpacking, biking, and bird-watching.

Pine Savannas: Ponderosa Pine

Distribution. The ponderosa pine savanna is a subset of the more extensive ponderosa pine forest. Ponderosa pine is the dominant tree species on over 16 million hectares of western North America (Arno et al. 1995, Barrett 1995), and 10% of this total, or about 1.6 million hectares, is estimated to be savanna. Ponderosa pine savannas occupy considerably less area than they did at the time of Anglo settlement (Gruell 1983, Covington et al. 1994, Arno et al. 1995), largely because of the combined effects of livestock grazing and fire exclusion (see chapter 4).

Savannas occur where ponderosa pine is virtually the only late-successional tree species and grasses are abundant in the understory (fig. 1.3). These conditions are best represented in the central and southern Rocky Mountains and the northern Sierra Madre Occidental, but they also occur in the most xeric regions of the maritime climatic regime of the Pacific Northwest. Savannas are particularly well represented throughout the Black Hills of South Dakota; adjacent to major drainages in Idaho, Washington, Oregon, and British Columbia; and in the foothills and broad ridges of the southern Rocky Mountains and the northern Sierra Madre Occidental.

Ponderosa pine is usually the lowest-elevation commercially valuable tree species throughout its geographic range (Steele 1987); it grows at elevations ranging from less than 300 m in the northern Rockies to over 2,200 m at some sites in the Sierra Madre Occidental. At the mesic upper limit, ponderosa pine savannas merge with closed-canopy conifer forests. These forests usually are dominated by ponderosa pine, but often one of several other conifers is present (e.g., Douglas-fir *[Pseudotsuga menziesii],*

Figure 1.3 Ponderosa pine savannas are dominated by ponderosa pine; other overstory species are uncommon.

white fir *[Abies concolor],* grand fir *[A. grandis],* western redcedar *[Thuja plicata]*). At the lower and drier limit, ponderosa pine savannas generally grade into bunchgrass-dominated grasslands or low shrublands at northern latitudes or grade into midheight shrublands, riparian forests, pinyon-juniper savannas, southwestern oak savannas, or bunchgrass- or shortgrass-dominated grasslands at southern latitudes (Linhart 1987, Steele 1987, Peet 1988).

Climate, Geology, and Soils. Average annual precipitation in ponderosa pine savannas typically is 250–500 mm (Pearson 1951, Schubert 1974, Clary 1987). Seasonality of precipitation varies from winter-dominant in the northern Rocky Mountains to summer-dominant in the Sierra Madre Occidental, with a nearly equal contribution of winter and summer precipitation in the central and southern Rockies (Pearson 1951). There is considerable variability in temperature regimes among sites occupied by ponderosa pine savannas, and the average frost-free period varies from 100 days in the northern Rocky Mountains to about 300 days in the southern Sierra Madre Occidental (Krajina 1965, Steele 1987).

Soils are derived from a wide variety of igneous, sedimentary, and metamorphic rocks, but the dominant geologic substrates are quartzite, argillite, schist, basalt, andesite, granite, and shale (Schubert 1974, Currie 1975, Wright and Bailey 1982). Soil depth ranges from a few centimeters to several meters, and soil texture varies from sandy to clayey (e.g., Pearson 1923, Rummell 1951, Potter and Green 1964, Schubert 1974, Steuter et al. 1990).

Vegetation Structure and Function. Ponderosa pine, an evergreen conifer, is often the only common woody plant. However, shade-tolerant species may establish in the understory, particularly in the absence of periodic fires. Subordinate species usually have persistent leaves, and most are conifers. There is a weak affinity between subordinate woody species and ponderosa pine taxonomy: species typical of maritime climates (e.g., Pacific madrone *[Arbutus menziesii],* incense-cedar *[Calocedrus decurrens],* tanoak *[Lithocarpus densiflorus]*) are found with Pacific ponderosa pine *(Pinus ponderosa* var. *ponderosa);* species associated with continental climates (e.g., white fir, curlleaf mountain mahogany *[Cercocarpus ledifolius],* Rocky Mountain juniper) co-occur with Rocky Mountain ponderosa pine *(P. ponderosa* var. *scopulorum);* and species with Madrean affinities (e.g., border pinyon, silverleaf oak *[Quercus hypoleucoides]*) are associated with Arizona ponderosa pine *(P. ponderosa* var. *arizonica).*

Dominant herbaceous species are perennial bunchgrasses and herbaceous dicots. Bunchgrasses are primarily native cool-season plants: the most common genera are fescue *(Festuca),* muhly *(Muhlenbergia),* and wheatgrass *(Agropyron)* (Krajina 1965, Clary 1987, Steele 1987, Johnson 1994). Two introduced C_3 grasses—the rhizomatous perennial Kentucky bluegrass *(Poa pratensis)* and the annual downy brome—commonly occur in the northern half of these savannas. South of the Mogollon Plateau in central Arizona and New Mexico, native warm-season grasses contribute up to 50% of herbaceous biomass: there, the primary genera are bluestem *(Andropogon)* and grama *(Bouteloua).*

Average annual stemwood production is 5–10 m^3/ha (Martin

1987, Van Hooser and Keegan 1987, Collins and Green 1988). Annual herbaceous production is about 1,700 kg/ha in northern savannas. Farther south, annual herbaceous production is generally lower (less than 1,200 kg/ha) and is more variable among sites and years (Clary 1987). Tree canopy cover and herbaceous production are inversely correlated (McPherson 1992a).

Land Use. Over 70% of the ponderosa pine savannas in the United States are administered by the USDA Forest Service, and other federal and state agencies are responsible for most of the remaining areas. In northern Mexico, *ejidos* are responsible for managing most of the ponderosa pine savannas, and the remaining Mexican lands are divided among federal and state trusts and individual owners (Whetten 1948, Felger and Wilson 1995).

As with the majority of North American savannas, livestock grazing represents the most widespread use of ponderosa pine savannas. Timber harvesting is also ubiquitous, and ponderosa pine is a valuable source of pulpwood, structural lumber, and specialty products (e.g., furniture, door frames, cabinets) (Blatner and Govett 1987). However, ponderosa pine savannas are characterized by low rates of postharvest establishment of trees and by low rates of tree growth relative to more mesic conifer forests. These factors contribute to the perception that ponderosa pine savannas are poorly suited for timber production (Clary 1987, Morgan 1987). Therefore, other uses for these savannas are becoming increasingly encouraged, particularly nonconsumptive recreational activities (e.g., Alexander 1986, Morgan 1987).

Easy public access, combined with public ownership and management targeted at multiple use, contributes to recreational opportunities that are common, widespread, and diverse. An abundance of big-game animals and upland game birds in ponderosa pine savannas makes hunting popular, and perennial water attracts anglers and white-water enthusiasts. Camping, hiking, and backpacking are also popular, particularly during the summer. High levels of recreational use contribute to significant tourism-based economies throughout the range of ponderosa pine savannas. Development of mountain homes has increased sharply on privately owned lands in the last two decades.

Pine Savannas: Longleaf Pine

Distribution. The longleaf pine savanna is a subset of the more extensive longleaf pine forest. Longleaf pine *(Pinus palustris)* is the dominant tree species on about 1.3 million hectares of southeastern North America (Landers et al. 1995), and a quarter of this total is estimated to be savanna. Longleaf pine dominated between 22 million and 35 million hectares at the time of Anglo settlement (Noss 1989, Frost 1993, Ware et al. 1993). Nearly all of the presettlement forest has been harvested, and reforestation programs and natural recruitment have not favored the re-establishment of longleaf pine.

The natural range of longleaf pine is a belt 150–250 km wide along the coastal plains of the Atlantic Ocean and the Gulf of Mexico from North Carolina to eastern Texas. Presettlement savannas were relatively continuous over much of the southeastern coastal plain below 75 m elevation and occurred on a variety of sites up to 600 m elevation (Mohr 1901). In contrast, current savannas are highly fragmented and are represented by small patches within agricultural fields, wetlands, pine plantations, or mixed mesophytic forests. These stands are generally less than 200 hectares in extent, and the largest old-growth stands cover less than 100 hectares. At the western extreme of its range, longleaf pine is bordered by agricultural fields and stands of loblolly and slash pine. These communities have largely replaced the southern end of the former midwestern oak savannas and the northeastern edge of mesquite savannas. Longleaf pine is bordered on the north by eastern deciduous or mixed mesophytic forest; at its southern and eastern limits, longleaf pine grades into slash pine forests or maritime communities.

Climate, Geology, and Soils. Warm, humid, subtropical climates typify longleaf pine savannas. Mean annual temperature varies from 17° to 23°C, and deviations of greater than 20°C from the mean are rare (Wahlenberg 1946). The frost-free period at the northern limit is about 200 days; it increases to 270–300 days near the southern limit or near the coast (Fowells 1965, Ware et al. 1993).

Mean annual precipitation varies from 1,100 to 1,600 mm, with a trend of increasing precipitation toward the coast. Precipitation is generally evenly distributed throughout the year, although occasional summer droughts occur in the western region and winter droughts occur farther east (Christensen 1988, Shoulders 1990).

At the time of Anglo settlement, longleaf pine occupied much of the coastal plain adjacent to the Piedmont Plateau. This region is characterized by a wedge of alluvial and marine sediments resting on a basement of Paleozoic and Precambrian rocks. Soils tend to be highly weathered. Longleaf pine savannas are found in soils that range from very well drained to poorly drained; in fact, longleaf pine tolerates a wide array of soil moisture conditions and is excluded only by prolonged soil saturation (Shoulders 1990). Coastal plain soils include typical cool-climate Spodosols in the northern coastal plain, Entisols on well-drained sands throughout the region, Inceptisols on alluvial plains, Alfisols in the southern region, and Ultisols (highly weathered soils) on poorly drained sites throughout the region (Christensen 1988).

Vegetation Structure and Function. Longleaf pine, an evergreen conifer, is often the only dominant overstory plant within longleaf pine savannas (fig. 1.4). Dominance is sometimes shared with shortleaf pine *(Pinus echinata).*

Subordinate species are winter-deciduous angiosperms, the composition of which varies with soil type. For example, turkey oak *(Quercus laevis)* is a common associate on ridge tops with poorly developed sandy soils. Farther downslope, soils tend to have well-developed clay horizons nearer the soil surface, and blackjack oak *(Q. marilandica),* sandhill post oak *(Q. margaretta),* and bluejack oak *(Q. incana)* are common. In addition, vines such as Virginia creeper *(Parthenocissus quinquefolia),* poison ivy *(Rhus radicans),* and the introduced kudzu *(Pueraria lobata)* are conspicuous on many sites. Post oak *(Q. stellata),* a dominant species in the southern portion of midwestern oak savannas, is commonly subordinate to longleaf pine in Louisiana and Texas (Bridges and Orzell 1989).

On relatively fertile and well-drained sites, shade-tolerant broadleaf angiosperms establish in the understory in the absence

Figure 1.4 Longleaf pine savannas are dominated by longleaf pine.

of periodic fires. As such, these stands are successional to southern mixed hardwood forest (Christensen 1988). In contrast, communities on xeric sandy soils and poorly drained areas display little evidence of compositional change, even when fires are excluded for long periods (Roberts and Oosting 1958, Veno 1976, Streng and Harcombe 1982, Abrahamson 1984). Thus, longleaf pine remains dominant on these sites, although density and cover increase to produce closed-canopy stands (Christensen 1988).

Dominant herbaceous plants are C_4 perennial bunchgrasses and a mixture of perennial and annual herbaceous dicots. Bluestems (*Andropogon* spp., *Schizachyrium* spp.) dominate upland sites in western longleaf pine savannas and fine-textured soils farther east. Historically, most upland sites were dominated by wiregrass (*Aristida stricta* in North Carolina and northern South Carolina, *A. beyrichiana* from southern South Carolina south throughout Florida and west to Mississippi; Peet 1993) (Frost et al. 1986). Wiregrass has been extirpated from most sites (Noss 1989) and is difficult to propagate (Clewell 1989, Duever 1989). Many sites formerly dominated by wiregrass are currently covered with Johnsongrass *(Sorghum halepense)* and cogongrass *(Imperata cylindrica)*, which are introduced C_4 grasses (Duever 1989).

Average annual wood production in longleaf pine savannas is 5–15 m^3/ha (Kelly and Bechtold 1990, Farrar 1993), and a trade-off between woody and herbaceous plants is present. Average annual herbaceous productivity varies from 200 to 3,000 kg/ha and is highly site-specific (Wolters 1981, Pearson et al. 1987, Stewart and Hurst 1987, Tanner 1987). Mesic sites are characterized by an extraordinary diversity of herbaceous plants, including numerous species of insectivorous plants. Plant species diversity on these sites is among the greatest in North America, with over 40 species of herbaceous plants per square meter (Frost et al. 1986, Peet and Allard 1993).

Land Use. About 70% of longleaf pine forests and savannas are privately owned, and the rest are managed by a variety of federal, state, and county agencies (Kelly and Bechtold 1990, Landers et al. 1995). One-fourth of the privately owned land is held by the forest industry, and the remainder—over half the total area—is managed by private landowners, mostly nonfarmers (Landers et al. 1995).

Primary contemporary land uses are livestock grazing, timber production, and production of game animals. Herbaceous production exceeds that of most savannas in western North America. In addition, interannual variability in herbaceous production is less than that of most western savannas. Therefore, livestock grazing is economically feasible and is widely practiced, especially on abandoned agricultural fields dominated by palatable grasses. The stemwood of longleaf pine is particularly valuable as a source of sawtimber and poles (Boyer and White 1990, Landers et al. 1995). However, these savannas are characterized by low tree establishment and low rates of tree growth compared to forests with higher tree densities or different dominant tree species (e.g., slash pine *[Pinus elliottii]*, loblolly pine *[P. taeda]*). As with ponderosa pine savannas, these factors contribute to the perception that longleaf pine savannas are poorly suited for timber production (Boyer and White 1990, Kelly and Bechtold 1990, Boyer 1993), so other uses are being sought (Landers et al. 1990, 1995). However, most comparisons of growth rate between longleaf pine and other pines are not valid for assessing differences in productivity between species: productivity rates have not been compared on similar sites, and longleaf pine typically occurs on soils that are infertile relative to those on which slash pine and loblolly pine occur.

Many game species, including white-tailed deer *(Odocoileus virginianus)*, wild turkey *(Meleagris gallopavo)*, bobwhite quail *(Colinus virginianus)*, and eastern cottontail *(Sylvilagus floridanus)*, are abundant in longleaf pine savannas (Engstrom 1993, Landers et al. 1995). Because most of the southeastern United States is privately owned, fee hunting is a potentially important land use (Landers et al. 1995). Current levels of fee hunting, like other recreational activities in longleaf pine savannas, are limited. However, some savannas, particularly those owned by wealthy industrialists, are managed as quail-hunting plantations. In fact, the history of longleaf pine savannas is inextricably linked to quail hunting, especially in the eastern half of the range of longleaf pine. Stoddard's pioneering work on fire-vegetation-wildlife interactions (e.g., Stoddard 1931) provides the basis for contemporary management of these plantations. A primary element of management for quail habitat is low-intensity prescribed fires every one to three years, a strategy that favors the maintenance of savanna physiognomy (Rosene 1969).

Longleaf pine needles ("pine straw") have been raked, baled, and sold as fertilizer or landscape mulch since at least 1922 (Mattoon 1922). The harvesting of pine straw is economically lucrative, with annual economic returns of $250–950/ha (Roise et al. 1991, Makus et al. 1994). However, the removal of pine straw may reduce tree growth (Jemison 1943, McLeod et al. 1979) and adversely affect nutrient cycles (Reinke et al. 1981), which has led conservationists to discourage the practice (Schafale and Weakley 1989). In addition, savannas represent unlikely sources of pine straw because pine needles are not particularly abundant on the soil surface beneath widely spaced savanna trees; grass-fueled surface fires consume the needles that are present; and grasses interfere with the harvest of needles.

Several sites are managed primarily for biodiversity and other conservation values. Examples include preserves owned by the Nature Conservancy, the Tall Timbers Research and Education Center, former plantations, and several state parks, national parks, and national forests (Ware et al. 1993). In addition, some of the largest blocks of longleaf pine savanna are found within military installations (e.g., Eglin Air Force Base); conservation is a high priority on these lands.

Oak Savannas: Californian

Distribution. The following summary is largely excerpted from a recent review of the history, ecology, and management of Californian oak savannas (Allen-Diaz et al. 1998). These savannas occur in about 3 million hectares near the western coast of the United States. The type occurs almost exclusively in California but is also represented by a few remnants in Oregon, Washington, southern British Columbia, and northern Baja California Norte.

Most savannas occur between 60 and 1,200 m elevation; the lowest-elevation savannas are found in the mountain ranges near the Pacific coast, and savannas at higher elevations are associated with the Sierra and Cascade foothills (Barbour 1987, Allen-Diaz et al. 1998). Savannas dominated by blue oak *(Quercus douglasii)* are extensively and continuously distributed in the foothills of the Sierra Nevada, and they encircle the Central Valley; elsewhere, savannas are fragmented and less extensive (White 1966, Griffin 1977).

Californian oak savannas generally merge into annual grasslands at lower elevations or with decreasing amounts of effective precipitation. Herbaceous species composition in these savannas resembles that of annual grasslands (Bartolome 1987). Occasionally, savannas merge with chaparral at low elevations (Griffin 1977, Parker 1994). Oak savannas grade into closed-canopy oak woodlands, ponderosa pine forests, or mixed conifer forests at higher elevations (Barbour 1987, Allen-Diaz and Holzman 1991, Borchert 1994).

Climate, Geology, and Soils. Californian oak savannas are characterized by a mediterranean climate, with cool, wet winters and hot, dry summers. Nearly all of the precipitation falls between October and April, and severe summer droughts average six months in duration (Pavlik et al. 1991). Average annual precipitation varies from 200 to 900 mm (Barbour 1987, Evett 1994). The annual frost-free period varies from about 300 days in the foothills of the northern Sierras and southern Cascades to over 360 days near the Pacific coast and in southern savannas.

Californian oak savannas are associated with a wide variety of geologic substrates and geomorphic surfaces: these savannas are found on essentially all parent materials (Allen et al. 1989, Allen-Diaz and Holzman 1991). Savannas dominated by blue oak, valley oak *(Q. lobata),* and coast live oak *(Q. agrifolia)* usually occur on sandstones, shales, or sedimentary materials, but granitic parent materials are found at the north and east sides of the Central Valley (Allen et al. 1989). Savannas dominated by interior live oak *(Q. wislizenii)* generally occur on metamorphic and deep igneous parent materials (Allen et al. 1989). Soils tend to be thin, rocky, and poorly developed, but a few savannas are associated with deep, loamy soils (Allen-Diaz and Holzman 1991, Allen-Diaz et al. 1998).

Vegetation Structure and Function. Common overstory species include six species of oak, ranging from evergreen (coast live oak, interior live oak) to drought-deciduous (Engelmann oak *[Quercus engelmannii]*) and winter-deciduous species (blue oak, valley oak, Oregon oak *[Q. garryana]*). The most widespread overstory species is blue oak, which is found throughout the foothills surrounding the Central Valley (Griffin 1977, Pavlik et al. 1991, Borchert 1994). Blue oak savannas nearly always include conspicuous representation by foothill pine *(Pinus sabiniana),* and interior live oak commonly occurs as a subordinate species (fig. 1.5). Interior live oak dominates a few savannas in the foothills (Griffin 1977), and Engelmann oak dominates relatively small, fragmented savannas in southern California and Baja California Norte, Mexico (Scott 1991). Engelmann oak codominates some southern savannas with coast live oak, and coast live oak is the dominant overstory species in savannas of the coast range from Mexico to northern California (Parker 1994). Valley oak is often found on deep soils in low hills and valleys, both in the Central Valley and in valleys of the coast ranges in California (Allen-Diaz et al. 1998). Oregon oak occupies similar habitats in the valleys of the coast range in Oregon, Washington, and southern British Columbia (Franklin and Dyrness 1988).

Figure 1.5 Californian oak savannas may be dominated by any of several oak species. The most common overstory dominant is blue oak. (Photograph courtesy of Mitch McClaran.)

Early researchers (e.g., White 1966, Griffin 1971, 1976) noted a conspicuous gap in the size-class distribution of overstory oaks: saplings were largely absent. These observations contributed to the development of a substantial body of knowledge regarding recruitment of oaks (e.g., Plumb 1980, Plumb and Pillsbury 1987, Standiford 1991). In particular, most recent work suggests that many seedlings emerge but that the seedlings usually must be protected from vertebrates to ensure their survival to the sapling stage (Adams et al. 1992, Hall et al. 1992). Oaks that exceed 2 m in height generally survive to maturity (Allen-Diaz et al. 1998).

Dominant understory plants include introduced C_3 annual grasses and herbaceous dicots with Mediterranean origins. Typical grass genera include brome *(Bromus)*, barley *(Hordeum)*, wild oats *(Avena)*, fescue, and rye *(Elymus, Lolium)*. Several genera of herbaceous dicots are generally represented, including filaree *(Erodium)*, clover *(Trifolium)*, and geranium *(Geranium)*.

Annual stemwood production averages 5–15 m^3/ha (McDonald 1980, Pillsbury et al. 1987, Pillsbury and Joseph 1991). Interannual variability in herbaceous production is high and varies greatly with site and precipitation (Bartolome 1987). For example, herbaceous production is about 1,000 kg/ha on relatively xeric sites or during dry years and about 5,000 kg/ha on mesic sites or during years with above-average precipitation (Kay and Leonard 1980, Bartolome 1987, Kay 1987, McClaran and Bartolome 1989a, Jackson et al. 1990, Ratliff et al. 1991).

Land Use. About 90% of Californian oak savannas are privately owned (FRRAP 1988, Allen-Diaz et al. 1998). Properties smaller than 80 hectares account for 98% of the owners but only 0.3% of the total extent. In contrast, properties larger than 2,000 hectares account for less than 0.1% of the owners but 88% of the total extent (Fortmann and Huntsinger 1987). Thus, a few owners are responsible for managing most oak savannas. Typically, small properties are characterized by nonconsumptive management, whereas large properties are managed as ranches for production of sheep and cattle (Huntsinger and Clawson 1989).

Although livestock grazing is the primary land use, other consumptive uses are common. For example, oaks are widely used in furniture and other specialty products (Hall and Allen 1980, Quarles 1987). However, the timber industry in the western United States has focused on conifers rather than angiosperms (Hall and Allen 1980, Smith 1987), so Californian oaks have never supported a large lumber industry (Quarles 1987). Fuelwood harvesting, hunting and fishing, and water harvesting are common management practices within oak savannas. The latter activity, which includes dam building and management activities designed to increase runoff and the subsequent storage of water, provides most of the water for the heavily populated areas along the Pacific coast.

Urban areas near the Pacific coast are among the most populous in North America. Urban development is constrained by the Pacific Ocean to the west, so recent development has occurred toward the east, into oak savannas. Thus, oak savannas are being displaced by urban expansion on a relatively large scale (see chapter 4). In fact, urbanization has been the major contributor to a recent, rapid decrease in the extent of Californian oak savannas (Bolsinger 1988, Huntsinger and Clawson 1989).

Oak Savannas: Southwestern

Distribution. The following summary is excerpted from a recent review of the history, ecology, and management of southwestern oak savannas (McClaran and McPherson 1998). These savannas generally are considered a subset of the more extensive evergreen oak woodlands, or encinal, most of which are too dense to be classified as savannas (Brown 1982, Rzedowski 1983). Southwestern oak savannas occur in about 1.5 million hectares of the Sierra Madre Occidental; about 90% of this area is located in northern Mexico, with the remainder in Arizona and New Mexico.

Southwestern oak savannas are restricted to the relatively low (1,100–2,200 m) and dry elevations where encinal grades into desert grassland or shrubland (White 1948, Gentry 1957, Rzedowski 1983, McPherson 1992b). In Arizona, New Mexico, and northern Sonora, these savannas generally are discontinuously distributed in the foothills of isolated mountain ranges. In Chihuahua, Durango, Sinaloa, and central-southern Sonora, these savannas are somewhat more continuously distributed across large regions with rolling topography (Rzedowski 1983).

Climate, Geology, and Soils. Summer daytime temperatures in southwestern oak savannas often exceed 35°C. The average frost-free period is about 200 days at the northern end of oak savannas and over 300 days at the southern end (Brown 1982).

Average annual precipitation varies from 350 to 600 mm and is bimodally distributed (White 1948, Gentry 1957). In the northern region of these savannas, about half of the annual precipitation is received in July, August, and September in the form of short-duration, high-intensity convective storms; most of the remaining precipitation occurs in the form of longer-lasting, low-intensity frontal storms that occur in winter. Some snow falls nearly every year, at least in the northern regions of these savannas. Farther south, there is a trend of decreasing winter precipitation, increasing summer precipitation, and warmer winter temperatures (Rzedowski 1983).

Geologic influences on oak savannas have not been described; the Sierra Madre Occidental is characterized by complex and

Figure 1.6 Southwestern oak savannas commonly occur in mountain foothills of the southwestern United States and northern Mexico. The most common overstory dominant is Emory oak.

poorly understood geology (Ortlieb and Roldan-Q. 1981). Southwestern oak savannas are associated with a large variety of geologic substrates and geomorphic surfaces, including granite, limestone, volcanic alluvial deposits, alluvial fans, and valley bottoms.

Most soils underlying oak savannas are shallow and rocky, although soil depth may exceed 6 m. Soils typically belong to the Aridisol order (Hendricks 1985), although some oak savannas occur on Mollisols (Medina 1987).

Vegetation Structure and Function. Common overstory plants are evergreen oaks. Emory oak *(Quercus emoryi)* is present nearly throughout the type in the United States and Mexico (fig. 1.6). In the United States and northern Mexico, it is associated with Mexican blue oak *(Q. oblongifolia),* Arizona white oak *(Q. arizonica),* and gray oak *(Q. grisea)* (Lasueur 1945, White 1948, Brown 1982, McPherson 1992b). Farther south in Chihuahua (Shreve 1939, Lasueur 1945) and Durango (Gentry 1957), associated oaks are Chihuahua oak *(Q. chihuahensis), Q. chuchuichupensis,* and *Q. santa-clarensis.* Cusi *(Q. albocincta)* is a common associate in Sinaloa (Gentry 1946).

The size-class distribution of oaks in U.S. savannas approximates a normal distribution (Sanchini 1981), with no evidence of major gaps in size- or age-classes (McClaran 1992) such as those reported for Californian oak savannas (McClaran and Bartolome 1989b). The absence of gaps in size-class distribution may be a result of the high propensity of southwestern oaks to resprout from the stump when cut or burned (Caprio and Zwolinski 1992, 1995) and for new plants to establish from acorns (Nyandiga and McPherson 1992).

Herbaceous species in these oak savannas include perennial bunchgrasses and several species of dicots (White 1948, Gentry 1957, Brown 1982, McPherson 1992b, McClaran et al. 1992, McPherson 1994). Introduced herbaceous species are uncommon in southwestern oak savannas, unlike other North American savannas. Many of the herbaceous species in these savannas are common in the desert grassland at lower elevations and in the denser encinal at higher elevations (Shreve 1939). Most graminoids are warm-season native plants, and the most common grasses are grama *(Bouteloua curtipendula, B. gracilis, B. hirsuta, B. radicosa),* lovegrass *(Eragrostis intermedia, E. mexicana),* and muhly *(Muhlenbergia emersleyi, M. longiligula).* Several species of perennial and annual herbaceous dicots emerge in early spring following cessation of freezing temperatures, and a second group emerges coincident with the summer monsoon (McPherson 1994).

Similar to other North American savannas, southwestern oak savannas are characterized by an inverse relationship between woody and herbaceous plants. Annual stemwood production averages 0.7–3.4 m^3/ha (Fowler and Ffolliott 1995). Annual herbaceous production is affected by soil type and slope: sandy-loam soils with 1–15% slope have relatively high production (1,100–2,800 kg/ha) compared to steep-sloped, rocky soils (220–350 kg/ha) (McClaran et al. 1992).

Land Use. Ejidos are responsible for managing about 70% of the oak savannas in northern Mexico, and the remaining areas are divided among federal and state trusts and individual owners (Whetten 1948, Felger and Wilson 1995). About 60% of the southwestern oak savannas in the United States are administered by the USDA Forest Service. Other federal agencies and the states of Arizona and New Mexico administer 20–25% of savannas, and 10–15% are privately owned.

Livestock grazing is the most common use of southwestern oak savannas. Livestock graze virtually all oak savannas in Mexico and about 75% of the southwestern oak savannas in the United States.

Recreational use of public and some private oak savannas has increased in the past 20 years, and it now represents the second most common use (McClaran and McPherson 1998). Deer and quail hunting have long been the dominant recreational uses, but increasing numbers of people visit these savannas to enjoy cooler temperatures, observe a variety of uncommon bird species, and generally escape the stress of urban life (McClaran et al. 1992). This increase in recreational use has spawned a significant ecotourism economy in southern Arizona.

Residential development has increased in savannas and adjacent desert grasslands in the last two decades. Some of the fastest-growing residential areas in the United States are near or within southwestern oak savannas (Bahre 1995).

The current quantity of fuelwood harvest on public lands in the United States is negligible (Bennett 1995), particularly compared to the harvesting that occurred during the mining era. Fuelwood was an important source of energy for home heating and cooking until the early 1940s in the United States (Bahre 1991). In Mexico, the quantity of fuelwood harvested has declined only since the 1970s because of the relatively recent availability of alternative fuels (Rzedowski 1983).

Finally, several cottage industries are supported by Mexican oak savannas. For example, fruits of native pepper plants (particularly chiltepine *[Capsicum annuum* var. *aviculare]*)and Emory oak are harvested and sold locally or exported. Beargrass *(Nolina microcarpa)* is fashioned into brooms or baskets, and many other plant materials are harvested for specialized uses (Huber 1992). The current supply of many of these products is greatly exceeded by the demand for them (Huber 1992, Felger and Wilson 1995).

Oak Savannas: Midwestern

Distribution. Midwestern oak savannas are discontinuously distributed from northern Minnesota to southeastern Texas. At the time of Anglo settlement, these savannas represented about 20 million hectares in a continuous band along the eastern edge of the Great Plains (Johnson 1986, Nuzzo 1986). Much of the area was plowed during the North American land rush, and the relatively productive soils are still used for crop production. Less fertile soils that were not plowed, or fields that were abandoned, have become closed-canopy woodlands during the last century. Thus, these savannas currently encompass only about 12,000 hectares, and their distribution is very fragmented (Johnson 1986, Nuzzo 1986, Smeins and Diamond 1986). The rarity of these savannas has caused them to be listed as a "globally imperiled" ecosystem (Heikens and Robertson 1994). Oak savannas are especially rare and fragmented north of Missouri; individual savannas in the upper Midwest rarely exceed 80 hectares in extent. These northern savannas are often considered distinct from their southern counterparts.

Midwestern oak savannas are transitional between oak forests to the east and prairies to the west. Because most of the adjacent prairies have been cultivated, oak savannas often form abrupt boundaries with fields of wheat, corn, or soybeans. In the absence of cultivation, oak savannas grade into prairies or pinyon-juniper savannas on the western boundary, mesquite savannas in central Texas, eastern deciduous forests on most of the eastern boundary, and southern pine forests in eastern Texas. The southern limit is marked by coastal plain communities, and adjacent natural communities to the north include pine barrens, pine forests, and tallgrass prairie.

Climate, Geology, and Soils. Average annual precipitation in midwestern oak savannas varies from about 650 to 1,200 mm, with trends of increasing precipitation from west to east and toward the Gulf of Mexico (Dyksterhuis 1948, Curtis 1959, Rice and Penfound 1959, Smeins and Diamond 1986). About half of the annual precipitation falls during the summer growing season. The frost-free period ranges from 100 days in Minnesota to 300 days in

Texas. The climate is sufficiently humid to support closed-canopy forest throughout the range of midwestern oak savannas. However, periodic severe droughts, such as those that occurred in the 1930s, 1950s, and 1980s, apparently kill enough woody plants to maintain savanna physiognomy.

There is no apparent correlation between the natural range of midwestern oak savannas and geological substrate or soil type (Bray 1960, Smeins and Diamond 1986). The few remaining examples of these oak savannas in the northern half of their natural range are restricted to relatively xeric shallow, rocky, or sandy soils; virtually all deep-soiled sites have been converted to agricultural fields (Nuzzo 1986). Similarly, southern oak savannas are underlain by sandy or loamy soils (usually Alfisols), with prairie or cultivated fields on more fertile clay soils (Smeins and Diamond 1986).

Vegetation Structure and Function. Dominant overstory plants are winter-deciduous oaks (fig. 1.7). In southern savannas, post oak is a characteristic species. Blackjack oak is also widespread but less abundant than post oak. Black hickory *(Carya texana),* black oak *(Q. velutina),* and white oak *(Q. alba)* may be locally abundant. Subordinate species include eastern redcedar—the easternmost member of pinyon-juniper savannas—and several winter-deciduous angiosperms: honey mesquite *(Prosopis glandulosa),* cedar elm *(Ulmus crassifolia),* common persimmon *(Diospyros virginiana),* and sugar hackberry *(Celtis laevigata)* (Smeins and Diamond 1986). In Missouri, this complement of southern species gives way to oak savannas with more northern affinities. Dominant overstory oaks are bur oak *(Quercus macrocarpa)* and northern pin oak *(Q. ellipsoidalis)* on most sites, with black oak and white oak codominant on other sites. Subordinate species are many and varied throughout the widespread, fragmented range of northern oak savannas (Nuzzo 1986).

Dominant herbaceous species include a diverse collection of graminoids and dicots typically associated with adjacent prairie. The most widespread dominant herbaceous plant is little bluestem *(Schizachyrium scoparium),* and other warm-season grasses are common on various sites. Subordinate species in the southern savannas include Indiangrass *(Sorghastrum nutans),* big bluestem

Figure 1.7 Midwestern oak savannas are scattered throughout the Great Plains; southern savannas such as this one commonly are dominated by post oak. (Photograph courtesy of Fred Smeins.)

(Andropogon gerardi), brownseed paspalum *(Paspalum plicatulum),* and introduced species such as bermudagrass *(Cynodon dactylon),* King Ranch bluestem *(Bothriochloa ischaemum* var. *songarica),* weeping lovegrass *(Eragrostis curvula),* and Johnsongrass (Smeins and Diamond 1986). Farther north, these species are joined by cool-season graminoids, including Kentucky bluegrass, sedges (*Carex* spp., *Cyperus* spp.), and needlegrasses (*Stipa* spp.) (Curtis 1959, Bray 1960, Heikens and Robertson 1995). Summer-flowering herbaceous dicots are common throughout both regions.

Average annual stemwood production is 1–2 m^3/ha (Ovington et al. 1963, Rosson 1994). Average annual herbaceous production is 2,000–5,000 kg/ha (Ovington et al. 1963, Wright and Bailey 1982).

Land Use. Slightly over half of the savannas are publicly owned by federal, state, county, and municipal agencies (Nuzzo 1986, Smeins and Diamond 1986). Many individual tracts occur in cemeteries, parks, or similar small, intensively managed parcels.

Thus, nonconsumptive recreational activities usually represent the primary use of these savannas.

An additional dominant land use is management for biodiversity and conservation values. At the national level, midwestern oak savannas have been identified as a critical area for preservation (Klopatek et al. 1979). At local levels, several state agencies and the Nature Conservancy have recognized midwestern oak savannas as threatened ecosystems and are managing them accordingly. Restoration is a primary goal of these organizations (e.g., Fralish et al. 1994). Livestock grazing is the only common consumptive use of midwestern oak savannas; grazing is largely restricted to savannas south of Missouri.

Mesquite Savannas

Distribution. Mesquite savannas cover about 35 million hectares in the southwestern United States and northern Mexico. Their spatial extent has increased since Anglo settlement as savannas have replaced former grasslands (see chapter 4). These savannas occur on flat or rolling topography where cultivation is limited by low precipitation or by rocky or shallow soils. Continuous stands of several thousand hectares are common. These large expanses often encompass patches of pinyon-juniper savanna on relatively coarse or rocky soils.

Mesquite savannas generally occur at elevations below 1,800 m in the United States but may extend to 2,500 m in Mexico (Brown 1982, Schmutz et al. 1991). These savannas often form an abrupt boundary with pine, pinyon-juniper, or oak woodlands at higher elevations. At lower elevations, they often merge into desert, desert grassland, or mixed-shrub vegetation; gallery forests or bosques of mesquite also are found in drainages within these communities. At the humid southern and eastern limits, mesquite savannas grade into subtropical thorn woodlands or pine forests.

Climate, Geology, and Soils. Mesquite savannas are associated with subtropical climates. Cold temperatures apparently constrain the northern and upper-elevational boundaries of mesquite (Bogusch

1951, Dahl 1982): the northern limit of mesquite corresponds with the average annual minimum temperature isotherm of −19.5°C (Fischer et al. 1959). Mesquite rarely occurs in areas with a frost-free period of less than 200 days (Dahl 1982).

Average annual precipitation generally ranges from 200 to 800 mm, with a trend of increasing precipitation from west to east. Mesquite persists on sites with less than 200 mm of annual precipitation, but grass cover becomes scattered or ephemeral. In western mesquite savannas, up to half the annual precipitation occurs in the winter months. Farther east in Texas and south into northern Mexico, summer thunderstorms contribute over 75% of the total annual precipitation (Brown 1982).

Mesquite savannas are underlain by a tremendous variety of soils and geologic substrates. Over a century ago, Havard (1884: 455) noted, "There is hardly any soil, if not habitually damp, in which mezquit *[sic]* cannot grow; no hill is too rocky or broken, no flat too sandy or saline, no dune too shifting . . . to entirely exclude it." However, savannas are limited to soils that are sufficiently deep to support a relatively continuous cover of grasses.

Vegetation Structure and Function. Mesquite, a deciduous angiosperm, is often the only dominant woody plant in mesquite savannas (fig. 1.8), although areas in southern Texas and adjacent northern Mexico are codominated by acacia (*Acacia* spp.). The Rocky Mountains and Sierra Madre Occidental generally form a boundary between velvet mesquite *(Prosopis velutina)* to the west and honey mesquite to the east. However, the two species intergrade, which contributes to an unclear understanding of *Prosopis* taxonomy (e.g., Benson and Darrow 1981). Subordinate woody species include a variety of winter- and summer-deciduous shrubs and succulents: acacia, creosote bush *(Larrea tridentata),* soaptree yucca *(Yucca elata),* tarbush *(Flourensia cernua),* prickly pear and cholla (*Opuntia* spp.), and snakeweed *(Gutierrezia sarothrae).*

Throughout most mesquite savannas, dominant herbaceous species are perennial C_4 grasses, most of which also dominate adjacent grasslands. Introduced grasses have become more abundant than native grasses on many sites within the last two decades. The most common introduced species are Lehmann lovegrass *(Eragrostis lehmanniana),* weeping lovegrass, buffelgrass

Figure 1.8 Mesquite savannas west of the Rocky Mountains, such as this one, are dominated by velvet mesquite.

(Pennisetum ciliare), and bermudagrass. Common native grass genera include numerous species of grama, three-awns *(Aristida),* lovegrass, muhly, and dropseed *(Sporobolus).* Tobosa *(Hilaria mutica),* curly-mesquite *(H. belangeri),* Arizona cottontop *(Digitaria californica),* and buffalograss *(Buchlöe dactyloides)* are also common on some sites. In northern Texas and New Mexico, the understory is dominated by C_3 grasses, particularly needlegrasses and squirreltail *(Elymus elymoides).*

Average stemwood production varies from less than 1 m^3/ha on xeric sites to over 20 m^3/ha on mesic sites. Annual herbaceous production varies from less than 300 kg/ha on sites with low annual precipitation to about 3,000 kg/ha on more mesic sites (S. C. Martin 1975, Tiedemann and Klemmedson 1977, Dahl et al. 1978, Dahl 1982, Heitschmidt et al. 1986, Brown and Archer 1989, Bush and Van Auken 1995).

Land Use. Over half of mesquite savannas occur on private lands, primarily in Texas. Most of the remainder is administered by *ejidos* in Mexico. The Bureau of Land Management, Department of Defense, National Park Service, Bureau of Indian Affairs, and state agencies oversee management of 10–20% of mesquite savannas.

Livestock grazing is the most common contemporary use of mesquite savannas. Over 90% of these savannas are grazed, primarily by cattle.

Historically, mesquite was an important source of fuelwood and charcoal. Mesquite was widely harvested for these uses until the early 1940s in the United States (Bahre 1991) and until the 1970s in Mexico (Rzedowski 1983). Harvest rates in Mexico have remained high, primarily to satisfy the demand for charcoal in the United States (Nabhan 1985). In addition, fuelwood and charcoal are still commonly used for heating and cooking in rural Mexico.

Mesquite wood is used for lumber, furniture, specialty products, and barbecue flavoring. Stands of large, widely spaced trees can be produced for the lumber industry by thinning and pruning (Meyer and Felker 1990, Cornejo-Oviedo et al. 1991). These stands produce up to 20 m^3/ha of lumber annually (Felker et al. 1990), which is comparable to the production of many other hardwoods (Cornejo-Oviedo et al. 1991). However, the production of mesquite for lumber is currently very limited; less than 50 m^3 of mesquite lumber is produced each year in the United States (Meyer and Felker 1990). Low volumetric shrinkage, high density, and high aesthetic value make mesquite a valuable wood for furniture, flooring, and crafts (Larson and Sodjoudee 1982, Weldon 1986). About 20,000 m^3 of small pieces and chips of mesquite wood are packaged and sold annually in retail outlets for the barbecue industry (Meyer and Felker 1990) at a wholesale value of about $3 million (Cornejo-Oviedo et al. 1992). Finally, mesquite trees are widely used for ornamental landscaping (Allworth-Ewalt 1982), with up to 10,000 wild trees transplanted annually (Meyer and Felker 1990).

Because most mesquite savannas are under private ownership, there are few widely accessible recreational activities on these savannas. Management of private lands for production of trophy big game is more lucrative on nearby pinyon-juniper woodlands than in mesquite savannas, so hunting is not an economically viable venture unless it is combined with other land uses. Recreational activities on public lands are similar to those on southwestern oak savannas: hunting, hiking, and bird-watching are common.

Savanna-Like Communities

Several other savanna-like communities occur in North America. These include grasslands with scattered yucca in the southwestern and midwestern United States and northern Mexico, sagebrush steppes centered in the Great Basin, aspen parklands and conifer barrens of the boreal region, and various other barrens and glades within the eastern deciduous forest. These communities are characterized by a mixture of woody plants and grasses, in proportions that are sometimes congruent with those of savannas. However, the overall physiognomy is more similar to grasslands, forests, or desert than to savannas. Thus, these vegetation types are not discussed in this book.

Summary

North American savannas represent an extensive, diverse, and important group of ecosystems. Different savannas are characterized by different climates, geologic substrates, soils, ownership patterns, and land uses, and this complexity represents a significant obstacle to the creation of a simple, integrated description. Nonetheless, the structure and function of seemingly disparate savannas are sufficiently similar to merit synthesis. The following chapters provide an overview of the patterns and processes shared by these ecosystems.

2

Overstory-Understory Interactions

You can observe a lot just by watching.
Yogi Berra

Do woody plants and grasses interact? If so, what are the expected outcomes from these interactions? Are interactions affected by life-history stages of plants? In an attempt to answer these questions, this chapter focuses on overstory-understory interactions at the spatial scale of individual plants. Fine-scale processes operate within the context of processes at the larger spatial scales discussed in chapter 3 (e.g., disturbance, climate, edaphic factors). The primary topics of this chapter are the effects of woody plants on the physical and biological environment (including herbaceous plants) and the effects of herbaceous plants on woody plants.

Effects of Woody Plants on the Understory Environment

Numerous effects, ranging from physical and chemical to biological, have been attributed to woody plants in savannas. Evidence for these effects usually is based on comparisons between areas beneath woody plants and adjacent grassland areas. However, such comparative research is hampered by weak inference: the "treatment" in these studies (i.e., the presence of a woody plant) may be confounded with pre-existing environmental conditions and therefore does not represent an experimental treatment per se. In other words, woody plants may establish and persist in areas that are physically, chemically, and biologically distinct relative to other areas on the landscape. For example, heterogeneous soil conditions (that cause, for example, concentrations of soil nutrients) may represent a pre-existing "natural" condition that is conducive to woody plant establishment and occupancy of a particular soil "patch." As such, microsites with high soil nutrient concentrations may represent "safe sites" (Harper et al. 1965) for woody plant establishment.

Distinguishing between correlation and cause and effect is facilitated by experimentation, including random assignment of treatments to experimental units (see chapter 7). Unfortunately, few experiments have been conducted in savannas. Thus, the results of comparative research are not necessarily reliable as mechanistic explanations (*sensu* Romesburg 1981). Nonetheless, the "effect" of woody plant canopies can be confidently assessed for some factors (e.g., light, precipitation distribution). Readers should interpret the following discussion of the results of comparative studies with considerable caution and an awareness of the limitations of the comparative approach.

Physical Environment

Physical and chemical properties beneath woody plant canopies often differ from those of the adjacent grassland. Effects of woody plants on microclimate and soil properties are often explained in terms of gradients: properties vary spatially outward from the woody plant bole, with a concentration of effect near the bole, a decline in effect with increasing distance from the bole, and no effect in interstitial areas. Among the physical and chemical effects attributed to woody plants are differences in light attenuation, soil temperature, precipitation distribution, soil moisture content, soil accumulation, and physical and chemical properties of soils.

Light Attenuation. Light attenuation by woody plant canopies has been widely documented. Photosynthetically active radiation is reduced by at least 85% beneath evergreen woody plants in pinyon-juniper (Smith and Stubbendieck 1990), Californian oak (Parker and Muller 1982), and southwestern oak savannas (Weltzin unpublished data) compared to adjacent grassland. Light

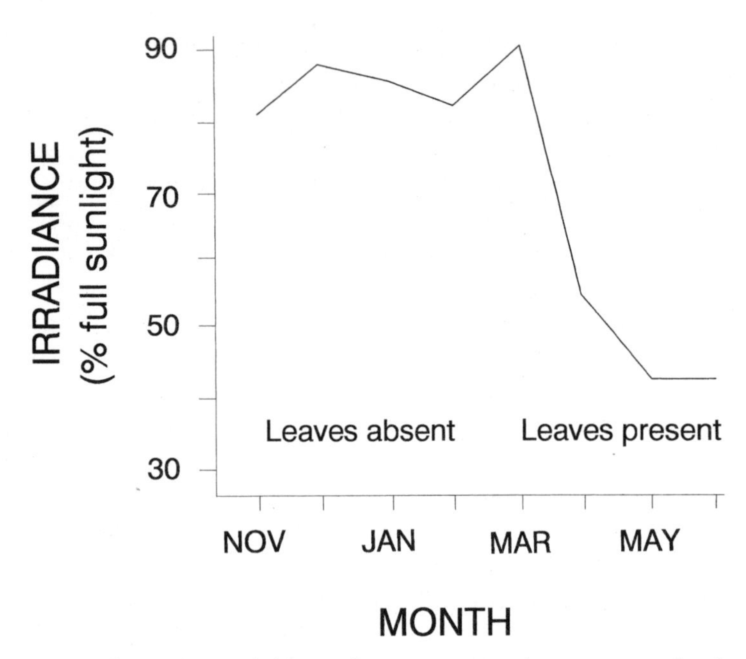

Figure 2.1 Irradiance beneath blue oak trees compared to open grassland in a Californian oak savanna. (Modified from Callaway et al. 1991.)

attenuation beneath deciduous woody plants depends strongly on leaf phenology (hence, season): irradiance beneath blue oak varied from 45% to 90% of that in the open, depending on the presence or absence of leaves (fig. 2.1; Callaway et al. 1991). In addition, light attenuation may vary considerably within a species, apparently as a function of site. For example, honey mesquite reduced radiation by 35% in southern Texas (Archer 1995b), 72% in central Texas (Bush and Van Auken 1990), 27% in northern Texas on an upland soil (Brock et al. 1978), and 82% in northern Texas on a more mesic, topographically lower soil. High-productivity sites (e.g., central Texas, mesic northern Texas) may support many leaf layers within a tree relative to low-productivity sites (e.g., southern Texas, upland northern Texas), thereby reducing light penetration beneath the canopy.

Light quality is little affected by woody plant canopies, although red light may be reduced beneath dense canopies, consistent with studies in other plant communities (Baskin and Baskin 1989, Sumrall et al. 1991). Shading of red light by a dense canopy may prevent stimulation of biologically active phytochrome (Bewley and Black 1982), thereby reducing emergence of some species (Baskin and Baskin 1989). Further, a high ratio of far red to red light may result in etiolation of certain understory plants.

Soil Temperature. Attenuation of incident radiation by woody plants generates understory thermal patterns distinct from those of grassland areas. The soil beneath woody plant canopies typically is cooler than grassland soil in spring and summer. In contrast, soil temperature beneath woody plants in fall and winter may be warmer than in the open, although this effect appears negligible for winter-deciduous woody species. In addition to reducing intra-annual variability in temperature, woody plants also reduce diurnal temperature fluctuations. These patterns have been observed in pinyon-juniper (Johnsen 1962), mesquite (Tiedemann and Klemmedson 1977, Fulbright et al. 1995), Californian oak (Holland 1973, Parker and Muller 1982, Callaway et al. 1991), and southwestern oak savannas (Haworth and McPherson 1995). Emory oak in southern Arizona demonstrates a pattern typical for evergreen species (fig. 2.2).

Precipitation Distribution. Woody plants produce local variations in precipitation patterns. This phenomenon has been well studied in closed-canopy forests (e.g., Cape et al. 1991 and references therein) and forest patches (e.g., Thurow et al. 1987, San Jose and Montes 1992) but not in savannas. It seems reasonable to assume that the influence of an isolated woody plant on precipitation distribution is different from that of a woody plant within a stand. For example, lateral movement of precipitation probably decreases with increasing density of woody plants. Unfortunately, there are insufficient data to permit evaluation of the effects of woody plant density and cover on precipitation distribution in any system.

Woody plants intercept incident precipitation and subsequently redistribute the moisture. Some precipitation is evapo-

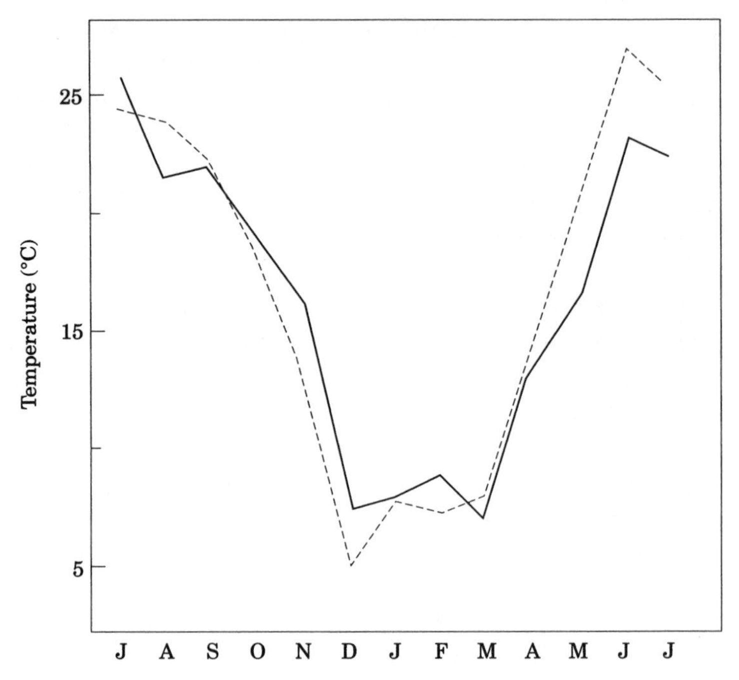

Figure 2.2 Mean monthly soil temperature at 15 cm depth beneath Emory oak trees (solid line) and in adjacent grassland areas (dashed line) in a southwestern oak savanna. (Adapted with permission from Haworth and McPherson 1995.)

rated from the canopy before it reaches the soil. The remaining precipitation passes through the canopy as throughfall or flows down the stem and is deposited at the tree base as stemflow.

Throughfall is influenced by characteristics of precipitation events and woody plant size and morphology, although these factors have rarely been quantified. For example, throughfall would be expected to increase with increased intensity and duration of precipitation. The interception of precipitation within woody plant canopies theoretically is related exponentially to event size (Helvey and Patric 1965). Atmospheric turbulence created by scattered woody plants and variable stem and leaf architecture within and among individual woody plants have contributed to highly variable measurements of throughfall in empirical studies (Young et al. 1984, Haworth and McPherson 1995). Nonetheless, throughfall in a pinyon-juniper savanna averaged about 20% (Young et al.

1984), and throughfall in a southwestern oak savanna ranged from about 25% in a 2 mm storm to 50% in a 13 mm storm to 100% in storms exceeding 35 mm (Haworth and McPherson 1995). Plant size also affects throughfall: large woody plants intercept more precipitation than do small woody plants.

Theoretically, stem and leaf architecture should have significant, predictable effects on the ability of a woody plant to intercept and store precipitation. Fissured or platy bark should impede water movement down the stem to a greater extent than smooth bark; rough, flat, or large leaves should enhance canopy storage more than smooth, angled, or small leaves; and V-shaped canopies should "funnel" water toward the plant base to a greater extent than plants with perpendicular branching (Barkman 1988). These ideas have not been evaluated empirically, possibly because of the strong effects of environmental conditions on precipitation distribution: wind speed, temperature, and humidity are among the factors that influence interception and throughfall.

Stemflow (the proportion of precipitation flowing down the bole of a woody plant) varies from 0% to 10% of gross precipitation (per unit area) in pinyon-juniper (Young et al. 1984, Young and Evans 1987) and southwestern oak savannas (Haworth and McPherson 1995). These are the only North American savannas for which data are available; these data are similar to those derived from closed-canopy forests (Tajchman et al. 1991), forest patches (Thurow et al. 1987), and deserts (Mauchamp and Janeua 1993). Small precipitation events create little or no stemflow because the woody plant canopy intercepts and stores the incident precipitation. This effect is greater for large woody plants than for small ones (Haworth and McPherson 1995).

Stemflow channels water directly to the base of the plant, where it can quickly reach relatively deep soil horizons. The quantity of nutrients in stemflow water may be greatly increased compared to that of ambient precipitation: leaf and stem exudates, atmospheric dust, bird excrement, invertebrate bodies, and spiderwebs are nutrient rich and contribute to nutrient-enriched stemflow. Thus, although the amount of stemflow (<10% of gross precipitation) seems relatively minor, stemflow may play an important role in establishing moisture and mineral gradients radiating out from the stem of the woody plant (Young et al. 1984,

Thurow et al. 1987, Young and Evans 1987). Gersper and Holowaychuk (1971:701) emphasized this factor in calling stemflow a "biohydrologic soil-forming factor." Stemflow is an important component of the water cycle of savannas and may help create mesic microenvironments beneath woody plants, as discussed below.

Soil Moisture Content. Woody plant–mediated changes in precipitation distribution, coupled with shade-induced reductions in evaporation and transpiration (Tiedemann and Klemmedson 1973), should result in varied soil moisture patterns in savannas. Other factors that may contribute to differential soil moisture beneath woody plants compared to the grassland zone include direct uptake of soil water by woody plant roots, and hydraulic lift (Caldwell 1990). The magnitude of each of these factors, along with their interactions, may partially explain soil moisture patterns in savannas. However, stemflow and throughfall are influenced by characteristics that vary with intensity and duration of precipitation events, species and size of woody plants, and environmental effects on water uptake (e.g., wind speed, temperature, relative humidity). Furthermore, the inherently high heterogeneity of soil physical characteristics (e.g., texture, porosity) may mask or override soil moisture effects attributable to woody plants.

Despite the number and complexity of factors that potentially affect soil moisture content in savannas, empirical evidence suggests that soil moisture content beneath woody plants rarely differs from that in adjacent grasslands (e.g., Tiedemann and Klemmedson 1977, Young et al. 1984, Engle et al. 1987, Jackson et al. 1990, Haworth and McPherson 1995). Thus, competing factors seem to negate one another. Occasionally, soils beneath woody plants are drier than adjacent grassland soils, particularly when woody plants are transpiring and grasses are dormant (e.g., savannas dominated by velvet mesquite [Tiedemann and Klemmedson 1977], western juniper [Young and Evans 1987], and eastern redcedar [Engle et al. 1987, Smith and Stubbendieck 1990]).

Soil Accumulation. Soil may accumulate beneath woody plants via animal inputs, by capture of airborne particles, or through accumulation and decay of plant material (e.g., leaves, stems).

Low levels of herbaceous cover between woody plants increase the potential for soil erosion and movement of plant material, thereby accentuating soil accumulation beneath woody plants and increasing elevation relative to interstices. Significant soil accumulation beneath woody plants has been reported in mesquite savannas (Virginia and Jarrell 1983, Archer et al. 1988) and is implied by substantially greater litter accumulation beneath trees than in adjacent open areas in most savannas (Tiedemann and Klemmedson 1977, Everett et al. 1983, Young et al. 1984, Schott and Pieper 1985, Young and Evans 1987, Jackson et al. 1990, McPherson et al. 1991).

Physical Properties of Soils. In addition to enhancing soil accumulation, woody plants may affect physical properties of soils. In particular, soil bulk density often differs between open areas and subcanopy areas. High bulk density negatively affects plant growth and development because the associated low porosity interferes with exchange of water and gases at the soil-plant interface. Low bulk density beneath woody plants has been attributed to accumulated litterfall and to the direct effects of roots on soil macroporosity (Kay 1987, Callaway et al. 1991). Woody plants may also attract fossorial mammals (e.g., pocket gophers, voles) that decrease bulk density by mixing the soil. Conversely, large animals seeking shade or forage may increase bulk density via trampling (Warren et al. 1986). Bulk density is generally lower beneath woody plants than in the nearby grassland (Tiedemann and Klemmedson 1973, Kay 1987, Callaway et al. 1991, Frost and Edinger 1991), and this effect decreases with increased depth. Bulk density increases after woody plants are removed, presumably because the effects of root growth, litterfall, and fossorial mammals are short-lived and are attenuated by other factors over time (Tiedemann and Klemmedson 1986, Kay 1987).

Chemical Properties of Soils. Woody plant–induced patterns of precipitation distribution, soil temperature, physical properties of soils, and animal activities may act individually or in combination to produce differential patterns of nutrient concentrations in savanna landscapes. Concentrations of nutrients beneath woody plants have been reported for nearly all North American savannas. In fact, "islands of fertility" (Garcia-Moya and McKell 1970, Schlesinger et al. 1990) are found throughout the world in plant communities with scattered woody plants. Scattered savanna trees may have acted as focal points for various cultural activities of Native Americans (e.g., eating, defecating, burying the dead). Leaves and branches accumulate dust and epiphytes (e.g., mosses, lichens), some of which are deposited beneath the canopy (Callaway and Nadkarni 1991). Epiphytes also increase the aboveground surface area of woody plants, further enhancing their ability to intercept nutrient-rich dust (Knops et al. 1996). The shade of woody plants attracts animals that subsequently deposit nutrients—in their bodies or feces—beneath trees (Archer 1995b). The broad lateral spread of woody plant roots relative to the region of litter deposition facilitates the relative concentration of nutrients beneath woody plant canopies (Barth 1980). Thus, the presence of scattered woody plants may contribute to nutrient depletion in the interstices and nutrient enrichment beneath woody plant canopies, and numerous natural and anthropogenic phenomena act as positive feedbacks to this process. Furthermore, soil chemical properties attributed to woody plants may persist for several decades after woody plants are removed (Klemmedson and Tiedemann 1986, Tiedemann and Klemmedson 1986, Barnes and Archer 1996).

Organic carbon contents are generally greater in subcanopy soils than in adjacent grassland soils (Potter and Green 1964, Tiedemann and Klemmedson 1973, Parker and Muller 1982, Klopatek 1987, Jackson et al. 1990, Frost and Edinger 1991, McPherson et al. 1991, McPherson et al. 1993, Barnes and Archer 1996). Carbon accumulation probably results from increased litterfall and root biomass associated with woody plants, and this accumulation may have important implications for the global carbon cycle. Specifically, recent increases in woody plant biomass in grasslands and savannas throughout the world may be an important carbon sink. Considering the large land area that is becoming increasingly dominated by woody plants (see chapter 4), increased carbon storage in the root mass and soil organic carbon beneath woody plants could be a significant phenomenon (McPherson et al. 1993).

Levels of total nitrogen (Tiedemann and Klemmedson 1973, Klopatek 1987, Jackson et al. 1990, Callaway et al. 1991, Frost and Edinger 1991, McPherson et al. 1993, Archer 1995b, Franco-Pizaña et al. 1995, Barnes and Archer 1996) and plant-available nitrogen (Parker and Muller 1982, Klopatek 1987, Parker and Billow 1987, Young and Evans 1987, Jackson et al. 1990, Callaway et al. 1991, Padien and Lajtha 1992, Franco-Pizaña et al. 1995) typically are greater in soils beneath woody plants than in the adjacent grassland (but see McPherson et al. 1991). Additionally, turnover rates of soil nitrogen are greater beneath woody plants than in the open (Jackson et al. 1990, Hibbard et al. 1993). Mesquites are the only dominant woody plants in North American savannas capable of biological nitrogen fixation; they fix 30–50% of their nitrogen (Johnson and Mayeux 1990).

Effects of woody plant canopies on other chemical constituents are more variable than effects on carbon and nitrogen. For example, the concentration of phosphorus generally does not differ between subcanopy and grassland areas (Tiedemann and Klemmedson 1973, Frost and Edinger 1991, McPherson et al. 1991, Franco-Pizaña et al. 1995), although soil phosphorus concentration was elevated beneath woody plants in some pinyon-juniper savannas (Klopatek 1987) and beneath blue oak trees in Californian oak savannas (Jackson et al. 1990, Callaway et al. 1991). Cations are generally present in higher concentrations beneath woody plants than in open areas (e.g., Tiedemann and Klemmedson 1973, Jackson et al. 1990, Frost and Edinger 1991, McPherson et al. 1991), although no studies report differences in all measured elements. Soil cations increased linearly with tree age beneath two-leaf pinyon, suggesting that accrual of cations in the soil involves simple release associated with litter decomposition (Barth 1980). Despite the acidity of leaves of many woody plants in savannas, soil surface pH generally does not differ between subcanopy and grassland areas (Potter and Green 1964, Tiedemann and Klemmedson 1973, Parker and Muller 1982, Jackson et al. 1990, Frost and Edinger 1991, McPherson et al. 1991; but see Klopatek 1987).

Biological Environment

Because of the number and magnitude of physical factors associated with woody plants, biological differences between subcanopy and grassland areas should be expected. However, heterogeneity of site characteristics (e.g., soil, geomorphology, disturbance) may mask expected differences (Haworth and McPherson 1995). In addition, because of reviewer bias, studies that fail to detect significant differences between subcanopy and grassland areas may be published at lower rates than those that do report differences (Mahoney 1976, Fagerstrom 1987).

Woody plants in savannas attract numerous animals, including humans. Moderate temperatures relative to the grassland, along with structural support for nests and perches, are likely attractants for mammals and birds. Many animals forage across a wide area and defecate or die in higher proportions beneath woody plants than in grassland, thereby transferring nutrients from the grassland to subcanopy locations. Birds that nest in woody plants represent the most visible example, but many native and domestic animals demonstrate similar behavioral patterns. Furthermore, differential faunal effects between woody and grassland areas probably are not restricted to aboveground: fossorial mammals may be more abundant or more active in one area than the other. The latter phenomenon has not been studied sufficiently to determine its prevalence or importance.

Woody plants may have positive, negative, or negligible effects on herbaceous biomass. Differential leaf phenology, canopy architecture, and rooting characteristics of species, as well as differences in the physical environment or disturbance patterns in different savannas, may account for disparate accounts of the effects of woody plants on herbs. For example, it is hypothesized that dense, evergreen canopies and shallow roots of woody plants in pinyon-juniper savannas may inhibit production of herbaceous biomass. In contrast, a relatively open canopy and deciduous leaves may ameliorate the microenvironment beneath mesquites and deciduous oaks during hot, dry months, thereby enhancing production of herbaceous plants on relatively xeric sites (Tiedemann and Klemmedson 1977, Menke 1989, Archer 1990). Dis-

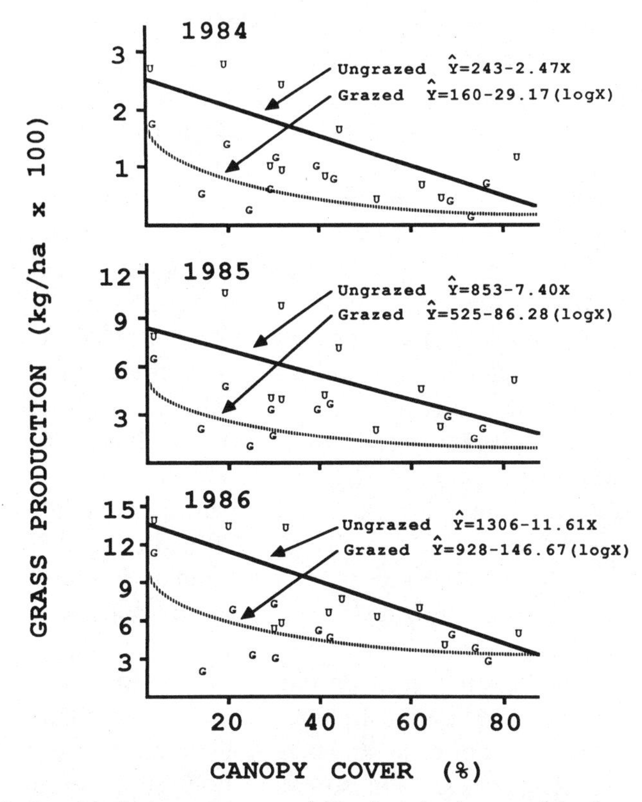

Figure 2.3 Relationship between cover of Pinchot juniper and herbaceous production on sites grazed (G) or ungrazed (U) by livestock. Precipitation was below average, average, and above average in 1984, 1985, and 1986, respectively. (Reprinted with permission from McPherson and Wright 1990a.)

turbances may reinforce or interact with the environment and growth form of woody plants. For example, dense canopies of some woody plants may protect herbaceous plants from livestock, thereby enhancing herbaceous production beneath canopies compared to the grassland (Vaitkus and Eddleman 1991).

Negative effects of woody plants on herbaceous production appear particularly common, and an inverse relationship between woody plants and herbaceous biomass has been reported for all North American savannas (e.g., see fig. 2.3): mesquite (Dahl et al.

1978, Wright and Van Dyne 1981), pinyon-juniper (Arnold 1964, Jameson 1966, 1967, Clary 1971, Everett et al. 1983, Schott and Pieper 1985, Engle et al. 1987, McPherson and Wright 1990a, Smith and Stubbendieck 1990, McPherson et al. 1991), Californian oak (Ratliff et al. 1991), southwestern oak (Haworth and McPherson 1994), ponderosa pine (McPherson 1992a), and longleaf pine (Wolters 1981). Decreased herbaceous biomass beneath woody plant canopies compared to open areas has been attributed to light attenuation, precipitation interception, chemical (i.e., allelopathy) and physical effects of litter accumulation, decreased availability of soil water or nutrients resulting from root activity of woody plants, and combinations of these factors.

Woody plants are also reported to have positive effects on herbaceous biomass (e.g., Holland 1973, 1980, Tiedemann and Klemmedson 1977, Archer 1990, Vaitkus and Eddleman 1991, Livingston et al. 1995, Callaway 1995). Increased herbaceous biomass beneath woody plants compared to grassland areas has been attributed to nutrient-rich subcanopy soils and microclimatic amelioration.

In some mesquite and longleaf pine savannas, woody plants may facilitate establishment of other woody plants. For example, honey mesquite facilitates establishment of Pinchot juniper in western Texas (McPherson et al. 1988). Pinchot juniper and oneseed juniper facilitate establishment of three woody species typically found in more mesic sites (Armentrout and Pieper 1988, McPherson et al. 1988). Honey mesquite also facilitates establishment of woody species in southern Texas, contributing to large, multispecies clusters of woody plants (Archer et al. 1988, Franco-Pizaña et al. 1995, Barnes and Archer 1996). Similarly, clusters of sclerophyllous evergreen oaks appear to represent loci of recruitment for several shrubs in longleaf pine savannas (Webber 1935, Laessle 1958, Guerin 1993). Mechanisms of facilitation have not been determined, and facilitation appears to be important for woody plants primarily during seedling establishment (Barnes and Archer 1996).

In addition to reports of negative and positive effects, a few studies have reported the absence of a relationship between woody plants and understory biomass (e.g., Brock et al. 1978, Heitschmidt et al. 1986, Jackson et al. 1990, Warren et al. 1996).

Blue oak, a winter-deciduous tree dominant in many Californian oak savannas, provides an excellent example of a well-studied woody plant in terms of canopy effects. Effects of blue oak canopy range from negative (Murphy and Crampton 1964, Murphy and Berry 1973, Kay and Leonard 1980, Kay 1987) to positive (Holland 1973, 1980, Holland and Morton 1980, Frost and McDougald 1989) and include reports of no measurable influence on understory biomass (Jackson et al. 1990) or cover (McClaran and Bartolome 1989a). Reported values of subcanopy herbaceous biomass range from less than 25% to more than 200% of adjacent grassland biomass. A plethora of data, generated by descriptive research at several locations, fueled a controversy over the effects of blue oak on herbs. Several hypotheses were forwarded to explain disparate results, and a series of experiments led to significant progress toward understanding these patterns (Callaway et al. 1991). These experiments indicated that herbaceous productivity is likely to be enhanced by nutrient inputs associated with the canopy but that the positive effects of increased nutrients may be negated or overridden by interference from shallow oak roots. The experiments did not address the probable continuum of effects induced by blue oak on herbs: these range from negative on mesic sites to positive on xeric sites (see references above). Nonetheless, variations in root morphology, which may be influenced by the precipitation regime, appear to explain the continuum of effects: shallow-rooted trees interfered with herbaceous production, whereas deep-rooted trees facilitated production (Callaway et al. 1991). Disparate effects of woody plants on herbaceous plants have also been attributed to variable root morphology in pinyon-juniper savannas (McPherson et al. 1991), suggesting that knowledge of root morphology of woody plants may be fundamentally important for predicting subcanopy herbaceous biomass.

Effects of woody plants on the diversity of understory plant species have been described in Californian oak and pinyon-juniper savannas with opposite results. Subcanopy herbaceous diversity was less than half that of adjacent grassland in an evergreen Californian oak savanna (Parker 1977). In pinyon-juniper savannas, increased species richness under woody plant canopies relative to the adjacent grassland has been attributed to pro-

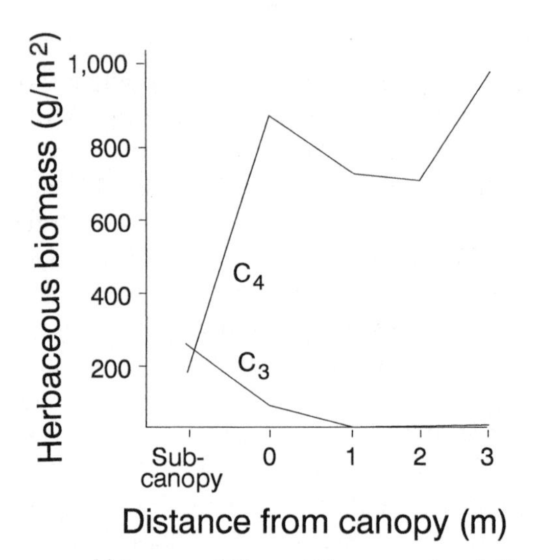

Figure 2.4 Aboveground biomass of C_3 and C_4 grasses in relation to distance from the canopy edge of Pinchot juniper trees in western Texas. (Data from the ungrazed site described in McPherson et al. 1991.)

tection from livestock in heavily grazed areas (McPherson and Wright 1990a, Vaitkus and Eddleman 1991). In addition, the presence of woody plants may create microenvironments conducive to occupation by a more diverse collection of plants than are found in grasslands. However, as woody plant cover increases to the point of forming closed-canopy stands, herbaceous diversity decreases (Koniak and Everett 1982).

Areas beneath woody plant canopies and the adjacent grassland often differ with respect to species composition (but see Haworth and McPherson 1994). For example, herbaceous plants with the C_3 photosynthetic pathway usually are more abundant beneath woody plant canopies than in the grassland, particularly in savannas dominated by C_4 grasses (fig. 2.4; see also Brock et al. 1978, Armentrout and Pieper 1988, McPherson et al. 1991, Archer 1995b). In an exceptional manipulative study, species composition of herbaceous plants associated with blocks of soil transplanted from coast live oak woodland to annual grassland, and vice versa, converged to resemble that of the "receptor" community within two years of transplanting (Marañon and Bartolome 1993). Artificial shade treatments produced responses similar to reciprocal soil transplants (Parker and Muller 1982, Marañon and Bartolome 1993). These studies indicate that shade exerts a greater influence over herbaceous species composition than do soil properties.

The number and variety of post hoc mechanisms invoked to explain observed overstory-understory patterns demonstrate considerable scientific creativity. However, experiments designed to evaluate these hypotheses have been scarce. This dearth of experiments is somewhat surprising, considering that experiments provide a powerful method to determine mechanisms underlying the observed patterns and that the spatial scale of the phenomenon is particularly conducive to experimentation (see chapter 7). For example, descriptive studies that spanned nearly four decades in Californian oak savannas fueled, rather than resolved, the controversy over effects of blue oak on herbaceous biomass, whereas a series of carefully designed manipulative experiments (Callaway et al. 1991, Knops et al. 1996) helped resolve the controversy. Experimental tests of other plausible hypotheses doubtless will provide additional insights.

In summary, woody plants in North American savannas have considerable influence on the physical and biological environments of the subcanopy. Their zone of influence tends to increase with increasing plant size (Johnsen 1962, Barth 1980, Everett et al. 1983, Haworth and McPherson 1994) but does not measurably extend beyond the canopy edge (Callaway et al. 1991, McPherson et al. 1991, Haworth and McPherson 1994, 1995). Effects are generally more obvious and consistent for physical than for biological factors. For example, differences in subcanopy and grassland zones are consistently directional in terms of light, soil temperature, precipitation distribution, and nutrient concentrations. In contrast, biomass, diversity, and species composition of herbaceous plants demonstrate less predictable patterns with respect to the presence of woody plants. The latter finding is at least partially attributable to the tendency for physical factors to interact or cancel out.

Effects of Herbaceous Plants on Woody Plants

The preceding section clearly demonstrates that woody plants can have substantial effects on the understory environment. However, the relationship between woody and herbaceous plants is not necessarily one-sided: considerable evidence indicates that understory plants interfere with recruitment of woody plants in savannas. In fact, herbaceous plants may interfere with woody plants in all life-history stages: germination, emergence, growth, survival, and reproduction. The following section reviews published effects of herbaceous plants on woody plants. The primary focus of this section is the "regeneration niche" (Grubb 1977) of woody plants, because other life-history stages are rarely studied. The regeneration niche includes the life-history stages from seed immigration through establishment. The concept is especially appropriate for studies of woody plant demography because early life-history stages usually represent the most critical constraints to woody plant abundance on a site.

Germination

Laboratory trials indicate that all common woody plants in North American savannas have inherently high germination rates (Young and Young 1992). However, germination is rarely isolated from emergence in field studies; when these events are isolated, germination in the field usually is lower than in controlled-environment facilities (e.g., Brown and Archer 1990, Bush and Van Auken 1990, Nyandiga and McPherson 1992, Young and Young 1992). The primary cause of this phenomenon is the difference in environmental conditions: higher levels of light and water are maintained in controlled-environment experiments than in field experiments. In field experiments, effects of herbaceous plants on woody plant germination do not appear to follow consistent trends, even within a species. For example, honey mesquite germination varied from 2% with herbaceous plants present to 19% in the absence of herbaceous plants on one site (Bush and Van Auken 1990), whereas herbaceous plants had negligible effects at a second, nearby site (Brown and Archer 1989).

Emergence

In contrast to the sparse and inconsistent data on germination, the effects of herbaceous plants on emergence of woody plants are well documented and nearly always negative. Low emergence rates may result from light attenuation, which can exceed 50% in dense stands of herbaceous plants (Bush and Van Auken 1990). In addition, dense herbaceous litter may suspend large reproductive structures (e.g., acorns, winged pine seeds) above the soil, thereby interfering with the ability of radicles to reach the soil surface (Platt et al. 1988b, Borchert et al. 1989). Also, herbaceous plants extract moisture from the soil; the resulting xeric environment reduces the likelihood of germination or emergence of woody plants (Pearson 1942, Scifres and Brock 1969, Griffin 1971, Adams et al. 1987, Gordon et al. 1989, Matsuda and McBride 1989, Adams et al. 1992).

Growth

Similar to emergence, growth of woody plant seedlings is constrained by herbaceous plants in all savannas for which data are available. Considerable research in controlled environments (e.g., Pessin and Chapman 1944, Smith et al. 1975, Nelson et al. 1985, Creighton et al. 1987, Van Auken and Bush 1987, 1988, 1990, Bush and Van Auken 1991, Koukoura and Menke 1995) and in the field (Chapman 1936b, Pessin 1938, 1939, 1944, Pearson 1942, Glendening and Paulsen 1955, Adams et al. 1987, 1992, Boyer 1989, Brown and Archer 1989, Gordon et al. 1989, Bush and Van Auken 1995) has demonstrated the combined effects of above- and belowground interference from herbaceous plants on growth of woody plant seedlings. The few field experiments designed to determine the relative importance of above- and belowground interference indicate that the latter is more important than the former in restricting growth of woody plant seedlings (Bush and Van Auken 1990, McPherson 1993). This finding suggests that herbaceous plants interfere with woody plant growth primarily via uptake of soil water or nutrients. In addition, different grass species exert differential effects on woody plants (Gordon et al. 1989, Rice et al. 1993).

Growth rates of adult woody plants also may be affected by herbaceous plants, although data regarding this premise are scant. The height of mesquite plants on a site grazed by livestock (hence, with reduced herbaceous interference) was nearly double the height of plants on an adjacent ungrazed site after eight years (Archer 1995a). However, differences in growth of mesquite plants between the two sites were not apparent until about five years after seedlings emerged. These findings are similar to data from southern Africa (Knoop and Walker 1985). Regrowth of Pinchot juniper was not affected by herbaceous cover during the three years following top removal (McPherson and Wright 1989). Given the ability of large woody plants to access deep sources of soil water (Heyward 1933, Lenhart 1934, Lewis and Burgy 1964, Flanagan et al. 1992), their obvious advantage in terms of light acquisition, and the slow growth rates of woody plants in savannas, it may be difficult to determine whether herbaceous plants significantly constrain woody plant growth in a consistent manner.

Survival

Survival of woody plant seedlings is critically affected by herbaceous plants. Most woody plant mortality attributed to interference from herbaceous plants occurs soon after emergence (Brown and Archer 1989, Gordon et al. 1989, Bush and Van Auken 1990, Welker and Menke 1990, McPherson 1993, Bush and Van Auken 1995), when woody plants and herbaceous plants directly compete for light and soil resources. Thus, gaps in the herbaceous layer may enable woody plant seedlings to access resources normally depleted by herbs. These gaps may be critically important to the survival of woody plant seedlings, particularly in systems characterized by high herbaceous biomass (e.g., Borchert et al. 1989, McPherson 1993, Bush and Van Auken 1995).

The ability of herbaceous plants to negatively affect growth and survival of woody plant seedlings increases with increasing herbaceous biomass, a pattern that parallels the influence of woody plants on herbs. In fact, a positive correlation between biomass and competitive impact may be a general characteristic of natural systems (Bonser and Reader 1995). Differences in herbaceous biomass may explain seemingly disparate effects of herbaceous plants on woody plant establishment. For example, reductions in herbaceous biomass below 600 g/m^2 did not affect honey mesquite survival (Brown and Archer 1989), but herbaceous biomass strongly influenced honey mesquite survival when aboveground herbaceous biomass was between 700 and 1,400 g/m^2 (Bush and Van Auken 1995).

Gap dynamics have been little studied in North American savannas, but it seems likely that gaps in herbaceous vegetation are important for woody plant establishment. Woody plants in savannas are capable of rapid root extension, particularly when herbaceous plants are removed. Thus, woody plant seedlings that establish in gaps can develop substantial root systems within a year or less. For example, seedlings in mesquite savannas produced tap roots 33 cm long in four weeks (velvet mesquite; Glendening and Paulsen 1955) and at least 40 cm long in four months (honey mesquite; Brown and Archer 1989); in the first year of establishment, roots of Emory oak in southwestern oak savannas may exceed 60 cm in length (McPherson 1993); and in the first five months of growth, roots of longleaf pine exceeded 30 cm in length (Lenhart 1934). Further, most grass roots are concentrated within the upper 30 cm of the soil (Pearson 1942, Schuster 1964, Weaver 1968, Brown and Archer 1990, Bush and Van Auken 1991). Thus, soil resource partitioning is facilitated and survival of woody plants is largely independent of subsequent development of the herbaceous layer. Variability in rates of gap formation may account for much of the observed variation in seedling establishment of woody plants between sites: relatively productive or frequently burned sites are characterized by continuous stands of grass and are therefore not conducive to woody plant establishment, whereas less productive sites (e.g., those with shallower soil or lower average annual precipitation) have considerable bare ground, which facilitates establishment of woody plant seedlings.

In addition to direct interference, herbaceous plants may indirectly affect mortality of woody plant seedlings. For example, juvenile woody plants are susceptible to fire and diseases, and the duration of the susceptible period may increase in the presence of herbaceous competitors. Woody plants become increasingly resistant to fire over time (Wright and Bailey 1982, Steuter and McPherson 1995), and longleaf pine in the grass stage is susceptible to brown spot *(Schirrhia acicola)* (Wakeley 1970). Nonetheless, longleaf pine seedlings "escape" the influence of herbaceous plants soon after they grow above the herbaceous plant canopy.

Reproduction

Herbaceous plants can also interfere with seed production of woody plants. Basal cover of herbaceous plants surrounding Pinchot juniper plants was negatively correlated with age of reproduction of Pinchot juniper (McPherson and Wright 1987), suggesting that reproductive age may be delayed in dense stands of herbs. Because attainment of minimum size is more important than attainment of age for reproductive maturity of woody plants (Young and Young 1992), effects of herbaceous plants on woody plant reproduction are largely indirect: by inhibiting seedling growth, herbaceous plants prolong the time required to attain the minimum size necessary for seed production by woody plants.

Summary

Herbaceous plants potentially exert considerable influence over establishment, growth, and reproduction of woody plants. However, the actual effects of herbaceous plants on woody plants depend on the abundance, phenology, morphology, and life-history characteristics of the interacting herbaceous and woody plants. For example, the ability of herbaceous plants to interfere with woody plants increases with increasing herbaceous biomass, which is consistent with the influence of woody plants on herbs. Interactions between woody and herbaceous plants are mediated by climate, soil, disturbance, and stochastic factors. These factors and their attendant effects on the interactions of woody and herbaceous plants are explored in the following chapter.

3

Savanna Genesis and Maintenance

To do science is to search for repeated patterns, not simply to accumulate facts.
R. H. MacArthur, Geographical Ecology

Why do savannas occur? If woody plants and grasses use and ostensibly compete for the same resources, and environmental conditions favor one life-form, then it seems reasonable to expect dominance by woody plants or by grasses instead of a mixture of the two. Therefore, do savannas simply represent temporally unstable transitions between grassland and woodland states? Or are they physiognomically stable at relatively long temporal scales? If savannas are stable, what factors are responsible for their genesis and maintenance? If not, what factors constrain their direction and rate of change?

Explanations for the existence of temperate savannas are evaluated in this chapter at a relatively coarse grain of resolution. Because savannas are recognizable only at relatively large spatial scales and because experimental manipulations are difficult to conduct at these scales, causal factors are difficult to determine. Nonetheless, the relative proportion of woody plants and grasses is controlled to a great extent by interactions between woody plants and grasses, and these interactions are clearly dependent on the environmental context (chapter 2). Therefore, this chapter focuses on herbivory, fire, climate, and edaphic factors, which are thought to affect the relative proportion of woody plants and grasses in systems capable of supporting both, or either, of the life-forms.

Herbivory

Differential herbivory of woody plants and grasses is one of the most important factors contributing to observed abundance of the two life-forms. For example, in most North American savannas, livestock grazing accelerates the rates of establishment and growth of woody plants to the detriment of herbaceous plants, particularly when environmental conditions that facilitate woody plant establishment coincide with the dormancy of herbs. In many Californian oak savannas, however, livestock grazing is detrimental to woody plants (Borchert et al. 1989, Hall et al. 1992). This section explores responses of savannas to herbivory and attempts to explain different outcomes of herbivory in different savanna types.

The variety and diversity of herbivores precludes general statements about their effects on vegetation. Savanna herbivores range from specialists that consume leaves of a single species to generalists that eat leaves, twigs, flowers, and fruits of many species. Further, herbivores range from small insects to large mammals. Even within a general category of herbivore (e.g., generalist large mammals), herbivory is variable among species (Stuth 1991). For example, cattle and sheep generally consume herbaceous plants preferentially over woody plants, whereas deer and goats consume a significant proportion of woody plants. This section begins by considering effects of domestic herbivores on savannas, because considerable research has been conducted on this topic. In contrast, relatively little research has been conducted on the importance of native herbivores in altering proportions of woody plants and grasses.

Livestock

Livestock grazing potentially increases the probability of woody plant recruitment in several ways (Archer 1994, McPherson and Weltzin 1997):

Livestock effectively disperse seeds of woody plants, primarily in their feces.

Removal of herbaceous biomass may increase the amount of light reaching the soil surface and thus increase germination and establishment of woody seedlings.

Reductions in grass leaf area with resultant reductions in root activity and biomass can (1) increase surficial soil moisture and thus enhance establishment and growth of shallow-rooted woody plant seedlings; (2) increase the amount of water percolating to deeper soil layers, thereby benefiting established woody species with deep root systems; (3) increase nutrient availability to woody plants; and (4) "release" suppressed populations of woody plants.

Grazing decreases basal area, increases mortality rates, and decreases seed production and seedling establishment of palatable grasses. Grazing may also increase susceptibility of grasses to other stresses (e.g., drought). Together, these factors increase rates of above- and belowground gap formation and thus increase the amount of area available for establishment of woody plant seedlings, especially in postdrought periods.

Grazing-induced shifts in herbaceous species composition may result in herbaceous assemblages less effective at competitively excluding woody plants or limiting their growth and seed production.

Reductions in fine-fuel biomass and continuity reduce fire frequency and intensity. Absence of fires accelerates shifts from grassland to woodland.

Woody species are often unpalatable relative to grasses and forbs and are thus not browsed with sufficient frequency or severity to limit their recruitment.

Loss of plant cover and consequent erosion reduce soil fertility, which favors N_2-fixing woody plants (e.g., mesquite) and growth forms tolerant of low nutrient conditions (e.g., evergreen shrubs).

Systematic eradication of prairie dogs, with the purpose of increasing herbaceous biomass for livestock consumption, may have removed a significant barrier to woody plant recruitment on many southwestern and midwestern landscapes.

Figure 3.1 Subtle differences in the history of livestock grazing may produce large differences in density and cover of woody plants. Both of these pinyon-juniper savannas have been grazed year-round by cattle for the last century. The pasture on the left side of the fence was grazed by sheep for 5 years, 15 years before the photograph was taken. The pasture on the right side of the fence was grazed by cattle during the same period.

The net effect of livestock grazing is selection for long-lived, unpalatable plants at the expense of short-lived palatable plants (fig. 3.1). Within this context, livestock grazing favors recruitment and growth of woody plants in most North American savannas. Californian oak savannas appear exceptional in this regard: oak establishment is adversely affected by livestock grazing, a phenomenon attributed to the unique plant phenology that is associated with mediterranean climates. In the mediterranean climate of Californian savannas, oak seedlings are among the few reliable sources of green and succulent forage during the summer, when most grasses are dead or dormant. Thus, oak seedling mortality increases as a result of increased consumption (Longhurst et al. 1979, Borchert et al. 1989) or trampling, as cattle become less discerning or more active in their quest for green forage or shade (Hall et al. 1992). Livestock in Californian oak savannas also consume a large number of acorns (Borchert et al. 1989), further contributing to reductions in oak recruitment.

Finally, activities associated with livestock production may indirectly affect interactions between herbaceous and woody plants. For example, livestock-induced reductions in herbaceous cover may contribute to changes in species composition of native herbivores, particularly small mammals (Glendening 1952, Humphrey 1958, Reynolds 1958, Buffington and Herbel 1965, Rice 1987). Subsequent shifts in populations of small mammals may significantly impact woody plant/grass ratios (see next section); these changes generally favor woody plant establishment. In addition, livestock producers exterminated prairie dogs (*Cynomys* spp.) throughout much of their original range in North America because the rodents were believed to compete with livestock for forage (Miller et al. 1990, 1994). Because prairie dogs (or other herbivores associated with prairie dog colonies) consume fruits of woody plants and remove tops of woody plant seedlings that do establish, their extirpation may have facilitated a transition from grassland to mesquite-dominated woodland around the turn of the century (Weltzin et al. 1997).

Native Herbivores

Native herbivores can influence the proportion of woody plants and herbaceous plants by disproportionally consuming or damaging more of one life-form than the other. For example, many common savanna birds (Grinnell 1936, Griffin 1976, Vander Wall and Balda 1977, 1981, McBride et al. 1991, Hubbard and McPherson 1997) and small mammals (Reynolds and Glendening 1949, 1950, Reynolds 1958, Griffin 1971, 1976, Cox et al. 1993) cache woody plant seeds at optimal germination depth, then fail to locate some seed caches. These native animals are believed by some researchers to be a primary mechanism of maintaining or increasing woody plant populations (e.g., Reynolds 1958, Vander Wall and Balda 1977, Cox et al. 1993). An exceptional long-term experiment illustrated that exclusion of kangaroo rats may contribute to an increased abundance of tall grasses in a shrub-dominated community in southeastern Arizona (Brown and Heske

1990). In contrast, several native rodents and lagomorphs in mesquite, oak, and longleaf pine savannas preferentially consume woody plant seeds (Brown et al. 1979, Cox et al. 1993) and repeatedly browse woody plants (Paulsen 1950, Campbell 1971, Griffin 1971, Wright and Bailey 1982, McBride et al. 1991, Weltzin et al. 1997). Thus, their exclusion may contribute to a reduction in the ratio of woody plants to grasses.

Invertebrate herbivory affects plant interactions in many ecosystems and demonstrably affects plant evolution in some North American savannas (e.g., Price 1991, Whitham et al. 1991). However, although much is known about the natural history of specific invertebrates, their influence on interactions of woody plants and grasses has not been determined in these savannas.

Effects of invertebrate herbivory on recruitment of woody plants have been described, in a limited manner, for oak savannas. Demographic studies indicate that invertebrate herbivory is a common source of mortality in recently emerged Emory oak seedlings in Arizona (McPherson 1993, Germaine 1997). Similarly, mesh exclosures designed to exclude invertebrates increase summer survival of blue oak and valley oak seedlings in California (Adams et al. 1992). However, the effects of physical factors such as reduced water stress are difficult to separate from the effects of exclosures on invertebrate herbivory because exclosures create shade that may reduce water stress; shade is critical for establishment of blue oak (Callaway 1992). Effects of invertebrates on woody plant recruitment in other North American savannas have not been described.

The presence of native herbivores has not prevented high rates of woody plant recruitment in many savanna systems. For example, southwestern oak savannas have displaced semidesert grasslands in southern Arizona with herbaceous plants and herbivores presumably present (McPherson et al. 1993, McClaran and McPherson 1998), and dramatic increases in mesquite density have been documented at sites inaccessible to livestock (Brown and Archer 1989). These and other similar studies were designed to investigate the net outcome of various mortality factors interacting on a site (i.e., they were not designed to partition out the relative contributions of various factors); they demonstrate that

Emory oak, honey mesquite, and velvet mesquite, like many other woody plants throughout the world, have overcome numerous constraints on establishment. Temporal and spatial variability in herbivory, caused by variations in populations of different animal species, doubtless contribute to substantial variability in native herbivores as sources of mortality (McPherson 1993, Germaine 1997). Thus, rates of woody plant establishment may be particularly high when populations of native herbivores are low.

At regional and larger spatial scales, abiotic phenomena are more important than herbivory and interference from herbaceous plants in constraining woody plant establishment. For example, post-Pleistocene patterns of woody plant establishment in pinyon-juniper (Jameson 1987, McPherson and Wright 1990b) and ponderosa pine savannas (Potter and Green 1964, Wells 1970, Savage 1991) support the hypothesis of strong abiotic control over vegetation physiognomy. At increasingly local scales of resolution, herbivory, herbaceous interference, edaphic factors, and disturbance become increasingly important determinants of vegetation patterns (Prentice 1986, Archer et al. 1995).

Fire

Fire has long been a pervasive and powerful force structuring plant communities in North America. Kimmins (1987) suggested that fire is as important a control on ecosystem structure and function as precipitation. Although the frequency, season, and behavior of fires influence plant communities, they are also shaped by local and regional vegetation patterns. As such, the long-term natural fire regime is probably more a consequence than a cause of vegetation patterns (Clark 1990).

Fire is reported to be a major factor in the maintenance of North American savannas (e.g., Wright and Bailey 1982, Platt et al. 1988b, Tester 1989, McPherson 1992a, Covington and Moore 1994b, Covington et al. 1994, Miller and Wigand 1994, Glitzenstein et al. 1995, Landers et al. 1995). Frequent fires have been linked to the development of longleaf pine savannas for over two centuries (e.g., Bartram 1791, Ruffin 1843) and with the development of midwestern oak savannas for nearly as long (e.g., Bourne 1820).

Plant response to fire varies widely among and within species and is affected by fire behavior and postfire physical and biological conditions. Fire behavior (size, intensity, and rate of spread) is influenced by (1) physical factors such as fuel conditions (e.g., moisture, total combustible material, continuity), weather conditions (e.g., wind speed, temperature, relative humidity), and topography (e.g., slope, aspect) and (2) biological factors such as plant morphology, physiology, and phenology, as well as community composition. Comprehensive reviews of the effects of fire on physical environments and individual plants, beyond the scope of this book, are provided by Wright and Bailey (1982), Krammes (1990), and Steuter and McPherson (1995). This section focuses on the response of savanna life-forms to fire.

Savanna grasses are well adapted to periodic fires. For example, seasonal aboveground production originates from buds or seeds near the soil surface, and nutrients move rapidly between above- and belowground biomass. Thus, high survival of extant individuals, along with recruitment of individuals from seed, facilitates rapid recolonization of burned areas. Total grass cover generally reaches or exceeds preburn levels within one year after a fire during the winter and within two to three years after a fire during the summer (Wright and Bailey 1982, Steuter and McPherson 1995).

Few juvenile woody plants in North America can survive even low-intensity fires. Two well-known exceptions are dominant savanna species: honey mesquite and longleaf pine. Both species establish belowground regenerative tissue very early in plant development. Honey mesquite produces a large tap root with numerous meristems at a young age, enabling most plants more than three years old to survive fires (Wright et al. 1976). Longleaf pine seedlings enter a grasslike morphological stage one to three years after emergence (the "grass" stage). During this stage, the apical meristem is protected by a cluster of long needles, and most plant resources are allocated to production of additional protective needles, thick stems, and an extensive root system, all of which offer further protection from fires (Pinchot 1899, Schwarz

1907, Mattoon 1922, Pessin 1938, 1944, Wahlenberg 1946, Bruce 1951, Grace and Platt 1995). Furthermore, longleaf pine seedlings in the "grass" stage are capable of resprouting following complete removal of the stem (e.g., Pessin 1938, Stone and Stone 1954, Garin 1958). Periodic fires during the "grass" stage apparently enhance seedling survival by removing dead leaves that harbor brown spot, which may otherwise kill the seedling (Verrall 1936, Wahlenberg 1946, Bruce 1951). After storing carbohydrates in the roots for several years in the "grass" stage, seedlings grow rapidly (more than a meter per year) for several years, which elevates the terminal bud beyond the reach of surface fires (Wahlenberg 1946, Bruce 1951). Even fires that occur before development of the "grass" stage are not catastrophic: nearly 20% of two-year-old seedlings in the pre-"grass" stage survive (Grace and Platt 1995).

Most woody plants that dominate North American savannas are resistant to surface fires when they are mature (Wright and Bailey 1982, Steuter and McPherson 1995). The thick, fissured bark of ponderosa pine and longleaf pine insulate the cambium. In addition, long pine needles whorled around buds absorb heat, thereby protecting the buds from lethal high temperatures. Most oaks in oak savannas also have thick, insulative bark. Mature individuals of most oaks and mesquite, and two juniper species (alligator juniper, Pinchot juniper), resprout from belowground buds if the aboveground portions of these plants are killed by fire. Thus, large woody plants in North American savannas are tolerant of frequent surface fires (Wright and Bailey 1982).

Geographic fragmentation associated with economic development (e.g., roads, housing developments), combined with efficient fire suppression, have essentially eliminated wildfires from savannas in the United States. Minimal fragmentation, poor transportation infrastructure, and inefficiency and indifference about fire suppression have contributed to fire regimes in Mexico that remain similar to historic regimes (Swetnam 1990, Fulé and Covington 1995). Wildfires are also common in Californian oak savannas because introduced annual grasses are sufficiently abundant to support periodic late-summer fires. Fires are particularly common in years when grass production exceeds the quantity of grass consumed by livestock.

The long-term absence of fire may produce dramatic changes

in community structure and function, particularly if soils do not limit woody plant establishment (cf. McAuliffe 1994). In the absence of fire, savannas may develop dense woody canopies that significantly reduce herbaceous production. Declining fine-fuel biomass over time reduces the probability of a surface fire, and the community changes from savanna to closed-canopy woodland (Heyward 1939, Brown 1982, Streng and Harcombe 1982, Archer 1989, Gilliam et al. 1993, Menges et al. 1993, Steuter and McPherson 1995). Crossing this "threshold" has serious implications for land management, as described below and in chapter 6: it represents a relatively permanent and significant change in vegetation (Archer 1989, Westoby et al. 1989).

Following several decades of fire suppression and recognition of subsequent changes in physiognomy, fires were reintroduced into some North American savannas. This trend began with the introduction of prescribed fires in longleaf pine savannas in the 1930s, based primarily on research by Chapman (1932, 1936a) and Stoddard (1931). Prescribed fires are widely used to manage longleaf pine savannas (Wade and Johansen 1986). Prescribed fires have not been universally accepted or widely used in other North American savannas, although periodic fires are used to reduce woody plants in mesquite and pinyon-juniper savannas (Wright 1974, Wright and Bailey 1982) and fire is used as a restoration tool in midwestern oak savannas (e.g., McClain et al. 1993, Fralish et al. 1994).

Once a former grassland or savanna is dominated by woody plants, herbaceous plants often are too scarce and discontinuous to support the spread of fire. In addition, oak and mesquite plants usually do not produce sufficient quantities of fine fuels at ground level to support the spread of fire, so these oak and mesquite savannas become essentially "fireproof" when they develop closed-canopy stands. Following canopy closure, fire alone cannot return the site to its earlier physiognomy or composition, and barring significant cultural inputs (e.g., herbicides or mechanical shrub control) the site is "permanently" occupied by woodland. A similar result is observed in most humid pine savannas, where the absence of fire contributes to increased cover of understory deciduous shrubs (Streng and Harcombe 1982, Gilliam et al. 1993, Menges et al. 1993), thereby reducing the flammability of these sites (Steuter and McPherson 1995). In contrast, increased density of pines or other highly flammable species (e.g., southern waxmyrtle *[Myrica cerifera],* holly [*Ilex* spp.]) may make some longleaf pine savannas more susceptible to high-intensity crown fires with increased time since the preceding fire (Christensen 1988). Fire exclusion also leads to overstory canopy closure in ponderosa pine savannas (Cooper 1960, Covington and Moore 1994a, 1994b), and these trees produce adequate fine fuel to support fire spread (Steuter and McPherson 1995). Thus, the long-term absence of fire in ponderosa pine savannas increases their susceptibility to high-intensity crown fires (Cooper 1960, Covington and Moore 1992, 1994a).

Canopy closure of pinyon-juniper savannas also can increase the probability of high-intensity crown fires. However, canopy closure rarely occurs in pinyon-juniper savannas, presumably because low woody plant densities are maintained by strong belowground interactions (see chapter 2). Neither herbaceous production nor leaf litter is sufficient to support the spread of surface fires in these stands. Thus, long-term fire exclusion in pinyon-juniper savannas tends to reduce the probability of future fire spread, primarily because fires do not spread between disjunct crowns except under extremely hot, dry, and windy conditions (Bruner and Klebenow 1979).

In summary, fire regimes exert considerable control over the proportion of woody plants and herbaceous plants in savannas. Fires tend to favor herbaceous plants at the expense of woody plants, although this depends on the behavior, season, and frequency of fires. High-intensity fires kill woody plants, with significant and persistent effects; by contrast, effects of low-intensity surface fires are minor and transient. In addition, frequent fires tend to favor herbaceous plants and widely spaced woody plants, whereas infrequent fires facilitate establishment and growth of woody plants (fig. 3.2). The virtual absence of fire has been a dominant feature of most North American savannas in the last century, with exceptions in Mexican and Californian oak savannas. Resultant effects in most savannas include increased abundance of woody plants and altered probability of future fires.

A

B

C

Figure 3.2 Fire frequency may strongly affect woody plant cover in savannas. These photographs of mesquite savannas were taken within 50 m of one another. (A) Mesquite plants are inconspicuous in areas burned 4–5 times per decade. (B) Mesquite plants are uncommon and scattered in areas burned 1–2 times per decade. (C) Mesquite plants dominate many areas that have not been burned for over 20 years. (Photographs reproduced by permission from McPherson 1995.)

Climate

Most research on climate and vegetation has been correlative and as such has limited predictive or explanatory power. Correlations between climatic variables and savanna distribution are unlikely to produce simple, single-factor models. For example, temperature regimes of North American savannas include coastal mediterranean (e.g., Californian oak in coastal mountain ranges), temperate continental (e.g., pinyon-juniper in the Great Basin region), and subtropical continental (e.g., mesquite in northern Mexico). Clearly, temperature alone cannot explain or predict the presence of savannas. Precipitation regimes associated with North American savannas are similarly variable and are therefore difficult to meaningfully correlate with savanna physiognomy.

Tropical and subtropical savannas are characterized by precipitation regimes that facilitate resource partitioning by woody plants and grasses (e.g., Walker et al. 1981, Walker and Noy-Meir 1982, Knoop and Walker 1985, Sala et al. 1989, Skarpe 1992). Walter's (1954, 1979) two-layer hypothesis of savanna stability suggests that shallow, fibrous-rooted grasses outcompete woody plants for shallow soil resources (e.g., water) when grasses are actively growing. In contrast, woody plants have exclusive access to resources deep in the soil profile (e.g., water that percolates to deeper soil layers when grasses are dormant). The two-layer hypothesis has been extended to North American savannas, where it has been used to predict the relative proportion of trees and grasses resulting from various scenarios of precipitation distribution (Neilson 1986, 1987, 1993, Neilson et al. 1992, Neilson and Marks 1994): summer precipitation is believed to support growth of herbaceous plants because these plants can exploit precipitation during the warm growing season; in contrast, winter precipitation percolates beyond the roots of dormant herbaceous plants and thus recharges deep soil layers accessed by woody plants. In North America, Walter's two-layer hypothesis appears to be applicable primarily to southwestern oak and mesquite savannas, many of which are characterized by a bimodal precipitation regime with peaks in summer and winter. However, the hypothesis has not been tested in these or other savannas.

Thus, as single factors, neither temperature nor precipitation can adequately explain the existence of North American savannas. However, the combined effects of these two factors may contribute to variations in the distribution and extent of North American savannas by controlling soil water balance (Griffin 1977, Neilson and Wullstein 1983, Eagleson and Segarra 1985, Neilson 1986, Rowlands 1993).

Temperature and precipitation patterns combine to produce a period of high plant water stress every summer in all North American savannas. For example, there is little or no precipitation when temperatures are greatest (June–September) in Californian oak, pinyon-juniper, and ponderosa pine savannas. Similarly, southwestern oak and mesquite savannas receive little or no precipitation during the hot foresummer (May, June), and summer evaporative demand exceeds precipitation in longleaf pine savannas (Christensen 1988). Thus, plant water stress, induced by seasonal patterns of temperature and precipitation, is a significant feature of all North American savannas. However, the magnitude of plant water stress varies considerably between seasons and years, thereby inhibiting generalizations about the overall effect of plant water stress on vegetation physiognomy.

Soil moisture is often considered the environmental factor that most constrains woody plant establishment in grasslands and savannas throughout the world (Skarpe 1992). Relatively long periods of high mean soil moisture may facilitate woody plant recruitment (e.g., Neilson 1986, Archer et al. 1988, McPherson and Wright 1990b, Turner 1990, Schmidt and Stubbendieck 1993). Subsequent growth of woody plants, along with additional recruitment, may transform savannas into closed-canopy woodlands (e.g., Burkhardt and Tisdale 1969, Young and Evans 1981, Jameson 1987, Archer 1989, McPherson and Wright 1990b). Once established, woody plants may persist for several decades regardless of climatic conditions (Neilson 1986, 1987, Archer 1990). In contrast, if low mean soil moistures restrict woody plant recruitment for periods exceeding the longevity of woody plants, savannas may become more open (Jameson 1987); however, the longevity of woody plants in North American savannas typically exceeds 200 years. Rapid recruitment in the presence of short-term increases in precipitation, coupled with the long life span of woody plants in North American savannas, contributes to relatively ephemeral effects of droughts on the woody component of these systems.

Climatic variability exerts considerable control over the structure and function of savannas, with impacts that may be similar to those caused by disturbances such as herbivory or fire. Interannual and decadal variability in precipitation and temperature are naturally high in North American savannas, particularly at local and regional scales. In fact, extreme climatic events may be more important than shifts in means (Katz and Brown 1992, Archer 1994). Thus, episodic climatic events may mask, confound, reinforce, or negate changes in vegetation distribution ostensibly attributed to changes in mean climatic condition. For example,

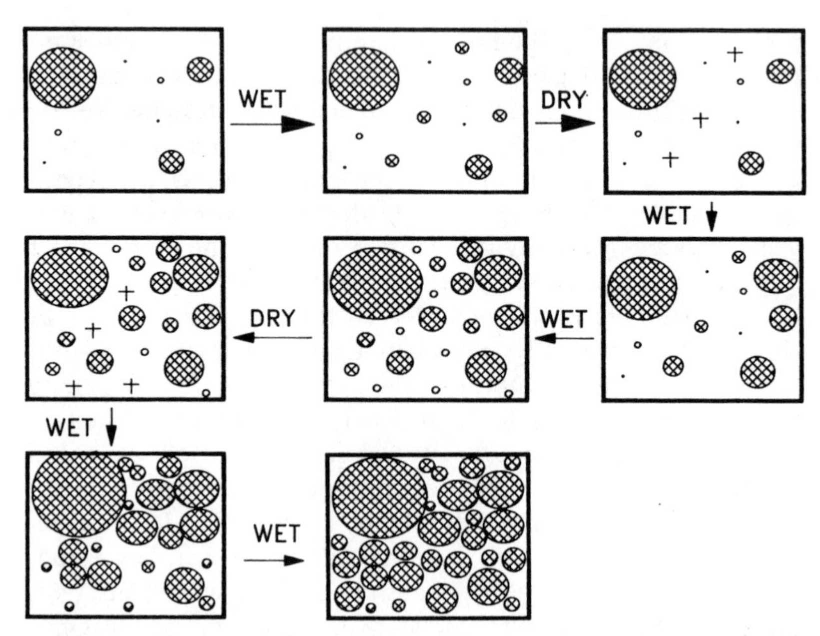

Figure 3.3 Short-term fluctuations in climate may produce persistent effects on vegetation. In a southern Texas mesquite savanna, the frequency and sequencing of dry periods influences the relative proportion of woody plant clusters (indicated with hatched circles). Dry years are characterized by mortality of clusters (indicated with +); years with average or above-average precipitation are characterized by formation of new clusters (indicated with a dot). Establishment of new clusters and expansion of existing clusters have led to substantial increases in woody plant cover on this site. Reprinted with permission from Archer 1990.

(1) drought may contribute to reductions in cover and productivity of grasses in a manner similar to livestock grazing (Buffington and Herbel 1965, Hennessey et al. 1983, Smeins and Merrill 1988); (2) mortality or top-kill of woody plants caused by drought or frost may have the same effects as fire; and (3) periodic droughts can be critically limiting to woody plant establishment in grasslands (see reviews in Wright and Bailey 1982, Archer 1994, McPherson and Weltzin 1997) and can contribute to decreased size of established plants (Archer et al. 1988, Archer 1990). Marshall (1957) observed considerable top-kill of trees in southwestern oak savannas at the end of a severe drought in the 1950s; this event was recorded in the photographic collection of Hastings

and Turner (1965). Parallel mortality was observed in ponderosa pine (Cooper 1960), pinyon-juniper (Betancourt et al. 1993), midwestern oak (Rice and Penfound 1959), and mesquite savannas (Carter 1964) during the same period. Pinyons and junipers have not recovered, but most of the oak trees have resprouted (Bahre 1991). Thus, vegetation changes resulting from a single climatic event (e.g., the drought of the 1950s) may be either ephemeral or persistent, depending on species-specific responses in different savannas.

"Pulses" of woody plant recruitment may result from relatively brief periods of high soil moisture (fig. 3.3). Periods of above-average precipitation may be particularly favorable for woody plant recruitment when they are preceded by severe drought: the drought reduces cover, density, and biomass of grasses, thereby minimizing interference and setting the stage for rapid increases in woody plant recruitment during the subsequent pluvial period (Smeins and Merrill 1988, Archer 1989, Archer and Smeins 1991). Thus, reductions in grasses and increases in woody plants combine to affect the relative proportion of woody plants and grasses.

Edaphic Factors

Savannas are underlain by a broad array of geologic substrates and soils (Buffington and Herbel 1965, Schubert 1974, Wright and Bailey 1982, Hawkins 1987, Glitzenstein et al. 1995, McClaran and McPherson 1998, Allen-Diaz et al. 1998). In general, density and cover of woody plants are positively correlated with annual rainfall and the proportion of coarse soil particles, which is consistent with a coarse-level classification of the world's savannas (Johnson and Tothill 1985).

At local scales, soils and vegetation are inextricably linked, and soil properties are strongly correlated with overlying vegetation (Cornelius et al. 1991). For example, in the absence of periodic fires, encroachment of woody angiosperms occurs more rapidly in longleaf pine savannas underlain by fine-textured soils than in those underlain by coarse-textured soils (Wells 1928,

Wells and Shunk 1931, Gilliam et al. 1993). Nonetheless, the paucity of research on soil-vegetation relationships at local scales prevents general statements about the influence of edaphic factors on savanna vegetation. A few systems have been studied in sufficient detail to permit researchers to draw limited conclusions, and these are discussed below.

Structure of woody vegetation is strongly correlated with soil properties in a southern Texas mesquite savanna (Archer 1995b). In this system, discrete shrub clusters have formed around mesquite plants within the last century (Archer et al. 1988, Archer 1989, 1990). Shrub cluster development has occurred much more slowly on soils with extensive argillic horizons (zones of clay enrichment which therefore are relatively impermeable to water and roots) than on nonargillic soils (Loomis 1989, Archer 1995b). The presence of an argillic horizon does not significantly constrain mesquite establishment, but it does reduce growth rates of individual plants and shrub clusters (Archer 1995b). Thus, it is predicted that the entire site will be occupied by closed-canopy woodland within the near future, regardless of underlying soils (Archer 1989, 1995b, Scanlan and Archer 1991).

Well-developed soil argillic horizons have been similarly implicated in vegetation structure in southern Arizona and southern New Mexico (McAuliffe 1994). Increased abundance of woody plants is retarded on sites with argillic horizons compared to adjacent sites with nonargillic soils, a difference attributed to slower growth and higher mortality of woody plants on the former soils (McAuliffe 1994).

Two potential mechanisms by which the presence of argillic horizons may suppress development of woodlands have been proposed (McAuliffe 1994): (1) argillic horizons reduce water availability to woody plants below thresholds necessary for survival in summer, and (2) water-impermeable argillic horizons result in perched water tables in the winter, hence contributing to woody plant mortality induced by physiological drought (i.e., flooding-induced oxygen deficiency). Available evidence supports the latter hypothesis (McAuliffe 1994), but neither has been tested.

Finally, interactions between soil type, land use, and fire regime affect vegetation development in savannas dominated by

Pinchot juniper. Consistent with most other savannas, juniper establishment is enhanced by livestock grazing and fire exclusion (e.g., McPherson et al. 1988). Soil movement and litter deposition protect the stem base of Pinchot juniper, which allows it to resprout from basal buds after fire (Steuter and Britton 1983, McPherson and Wright 1989). Because of differences in soil movement and site productivity, the basal portion of plants is protected more quickly on deep-soiled, moderately sloped sites than on shallow-soiled, flat sites (about 10 vs. 20 years, respectively). Thus, fires must occur approximately twice as frequently on deep-soiled sites as on shallow-soiled sites to maintain these savannas (Steuter and McPherson 1995).

Summary

Native and domestic animals, fire, climate, and soils are potentially important contributors to the genesis and maintenance of savannas. Clearly, no single factor exerts primary control over the relative proportion of woody plant and grasses in all North American savannas. Many factors and their interactions are important, and their roles vary spatially and temporally. Understanding how these factors interact is fundamental to interpreting and predicting responses of savannas to changes in climate, disturbance regimes, or management. Unfortunately, no savanna has been studied in sufficient detail to determine the relative importance of various phenomena in controlling the relative proportion of woody plants and grasses, much less to accurately predict response to perturbations.

A well-studied mesquite savanna in southern Texas illustrates the multiplicity of factors that influence community structure (Archer 1995b). In this system, livestock grazing (possibly in association with drought) apparently reduced grass cover in a former grassland. Reduced grass cover contributed to a reduction in fire frequency. In addition to reducing grass cover, cattle facilitated dispersal of mesquite by eating pods and defecating the viable seeds. Mesquite plants ameliorated the microenvironment and served as a focal point for native seed dispersers, thereby cata-

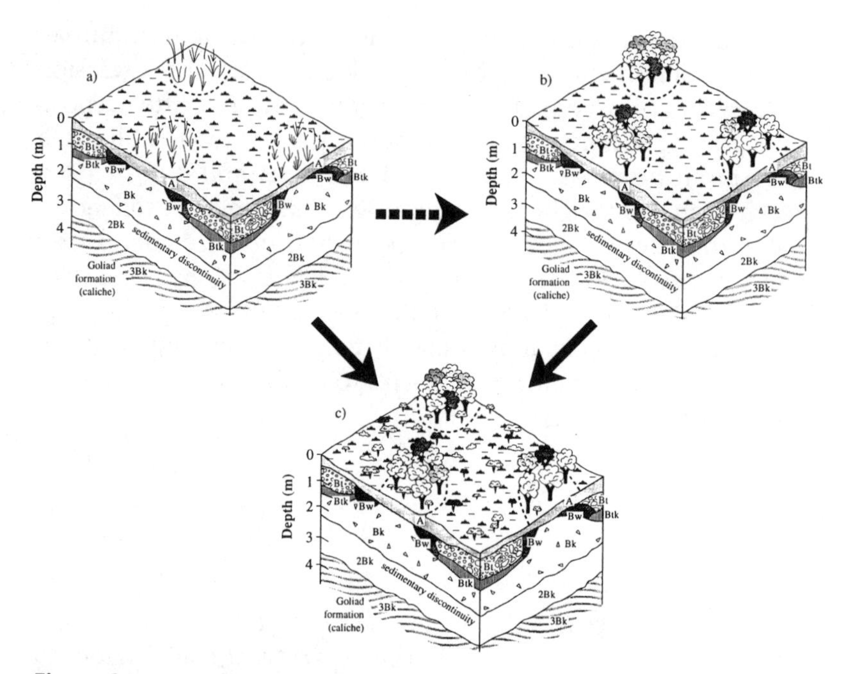

Figure 3.4 Hypothesized patterns of woody plant recruitment in southern Texas. Each panel depicts characteristic soil horizonation on the site. (A) Presettlement vegetation with short to midheight grasses dominant where the laterally extensive argillic (Bt) horizon is present; midheight and tall grasses dominate nonargillic (Bk) inclusions. (B) Presettlement vegetation similar to view A except that mesquite groves occur on nonargillic inclusions instead of grasses. (C) Present-day vegetation with mesquite groves on nonargillic inclusions and small, discrete clusters of shrubs dispersed among herbaceous vegetation where the argillic horizon is present. Arrows suggest pathways of directional change; available data suggest that a direct change from A to C is most likely (Archer 1995b). (Reprinted with permission from Archer 1995b.)

lyzing the development of multispecies clusters of woody plants. Mesquite growth and shrub-cluster development occurred most rapidly on nonargillic inclusions (see preceding section), but clusters have coalesced into a closed-canopy woodland virtually throughout the site. Thus, interactions between physical and biological factors, as well as historical events, have initiated and facilitated a conversion from grassland to savanna to dense woodland within a 150-year period (fig. 3.4).

Even a complete understanding of savanna genesis and maintenance at a particular site may contribute little to identifying and understanding important processes in other savannas. Nonetheless, general principles have emerged with respect to the response of dominant life-forms to various phenomena (e.g., livestock grazing, fire, precipitation pulses). These principles can be combined with knowledge of the natural histories of individual species to make coarse-level predictions about the response of a particular savanna to changes in land use or climate. Such predictions are developed in greater detail in chapter 5, after a review of past changes in savanna vegetation (chapter 4).

4

Historical Changes

There is something fascinating about science. One gets such wholesale returns of conjecture out of such a trifling investment of fact.
Mark Twain, Life on the Mississippi

Vegetation change in North America has been addressed by a large body of literature including several recent reviews (e.g., Humphrey 1987, Archer and Smeins 1991, Bahre 1991, Dick-Peddie 1993, Covington et al. 1994, McPherson and Weltzin 1997). Taken collectively, these studies provide a reasonably clear picture of past changes in savanna vegetation. These changes, and their potential causes, are summarized in this chapter. However, the relative importance of various mechanisms underlying past changes is generally unknown, at least in part because of methodological shortcomings associated with specific retrospective techniques described in this chapter. Because retrospective approaches are clearly inadequate for elucidating mechanisms of vegetation change, the chapter concludes with a discussion of the limitations of these approaches for evaluating current and future management activities.

Past Changes

Changes in management goals have contributed to changes in type and extent of land uses, particularly on public lands. For example, the pre-1930 goal of maximizing livestock production to feed a growing nation has yielded to goals based on consumptive and nonconsumptive multiple uses and maintenance of values (e.g., aesthetic, cultural, historic) that have increasingly been viewed as important. Management practices have changed to reflect these goals: livestock grazing is closely regulated, and other uses (e.g., ecotourism) have started to be factored into resource management plans. Altered management practices have influenced the extent, structure, and function of savannas via direct and indirect effects. Specifically, savannas have been characterized by urbanization, agricultural expansion, invasion of introduced plant species, and altered distribution and abundance of woody plants.

Urbanization and Agricultural Expansion

At local scales, urbanization and agricultural expansion have contributed to reductions in the extent of all savannas (fig. 4.1). These reductions largely coincided with Anglo settlement. Therefore, they began to substantially affect longleaf pine savannas during the 1700s. During the mid 1800s, land and gold began to attract large numbers of people to midwestern and Californian oak savannas, respectively. Ponderosa pine, southwestern oak, and mesquite savannas are located in relatively arid environments with soils poorly suited for agriculture; as a result, urbanization and agricultural expansion have proceeded more slowly in these savannas than in the other major North American savannas. Nonetheless, the "baby boom" after World War II and recent advances in technology and communication have contributed to the establishment of large urban centers in all North American savannas.

Urbanization is not a simple process by which land is converted at once from non-use to substrate for skyscrapers. Rather, land conversion begins with development of improved vehicular access. These roads are generally intended to facilitate wood harvesting, mineral extraction, and recreation. This phase has been completed in most areas of the United States, but oak and pinyon-juniper savannas in Mexico still have relatively few roads, most of which are poorly maintained. A population influx usually accompanies the improved transportation infrastructure, and land may be transferred to private ownership and then divided into smaller parcels. Land values tend to increase, with subsequent division of land into even smaller parcels. Eventually, industries based on natural resources (e.g., logging, mining, livestock grazing) may

Figure 4.1 Urbanization and agricultural expansion have contributed to reductions in extent of all North American savannas. (Photograph courtesy of Lynn Huntsinger.)

be partially or wholly replaced by businesses based on human resources (e.g., tourism, financial services, construction). This leads to further population influx (FRRAP 1988).

Urbanization has been the principal cause of a rapid recent decrease in the extent of Californian oak savannas (Bolsinger 1988, Huntsinger and Clawson 1989). In less than 150 years, California changed from a sparsely populated frontier to the most populous state in the United States (Williams 1991) at a rate greater than that of any other developing area in the world (FRRAP 1988). The foothills of the coast range and the Sierra Nevada, which were dominated by oak savannas before Anglo settlement, have been primary areas of land-use change and urbanization. Initially, urbanization was concentrated along the western coast and impacted savannas of the coast ranges; since 1950, development has shifted to the foothills of the Sierra Nevada and has primarily influenced blue oak savannas (FRRAP 1988).

At a regional scale of resolution, agricultural expansion was the primary reason for the virtual loss of midwestern oak savannas (Auclair 1976, Nuzzo 1986). In addition, agricultural expansion removed most presettlement longleaf pine savannas (Wahlenberg 1946, Landers at al. 1995). Poor sustainability of crop production contributed to the abandonment of many agricultural fields in the southeastern United States. However, longleaf pine savannas rarely re-established in abandoned fields (Landers et al. 1995); rather, pine forests dominated by loblolly pine and shortleaf pine have reoccupied many of these former longleaf pine savannas (Turner and Ruscher 1988). Loblolly pine and shortleaf pine forests usually have closed canopies and are managed primarily for timber production.

Introduction of Plant Species

Since Anglo settlement, introduced species have become established throughout most savannas. Only ponderosa pine and southwestern oak savannas have escaped large-scale encroachment by introduced plants, and they appear susceptible to invasion by downy brome and Lehmann lovegrass, respectively (see chapter 5).

Nonnative perennial herbs were widely introduced after about 1930 for purposes of erosion control and increased forage production for livestock (OTA 1993). Several species spread from seeded areas and have become regionally dominant. Examples include buffelgrass and bermudagrass in mesquite savannas, crested wheatgrass *(Agropyron cristatum)* in pinyon-juniper savannas, several species of lovegrass in both mesquite and pinyon-juniper savannas, and kudzu and Johnson grass in longleaf pine communities.

In addition to purposeful introductions, there have been several accidental introductions of nonnative species. For example, the understory of Californian oak savannas is dominated by annual grasses inadvertently introduced from the Mediterranean region (see chapter 1). Similarly, nonnative winter annual grasses (e.g., downy brome and medusahead) are well represented in pinyon-juniper savannas. Leafy spurge *(Euphorbia esula),* a perennial herb inadvertently introduced in the midwestern United States, is now widespread in midwestern oak savannas and pinyon-juniper savannas of the Great Plains.

Introduced species have altered the structure and function of many North American ecosystems (OTA 1993). For example, nonnative plants interfere with growth and survival of native grasses and woody plants (e.g., Harris 1967, 1977, Gordon et al. 1989, Aguirre and Johnson 1991, Gordon and Rice 1993). In addition, the associated increased production may contribute to increased fire frequency (see chapter 5). Thus, management goals that include maintenance of presettlement structure and function may require extirpation of introduced species. However, introduced species are difficult to extirpate because they are well adapted to areas they currently dominate and because they may establish and spread with little or no disturbance (D'Antonio and Vitousek 1992).

Altered Distribution and Abundance of Woody Plants

The distributions of most dominant woody plants in North American savannas have increased in areal extent since Anglo settlement, primarily through encroachment into former grasslands (fig. 4.2). Range expansions have been described for mesquite (Brown 1950, Hastings and Turner 1965, Hennessey et al. 1983, Archer et al. 1988, McPherson et al. 1988, 1993, Tieszen and Archer 1990), ponderosa pine (Potter and Green 1964, Progulske 1974, Bock and Bock 1984, Fisher et al. 1987, Steinauer and Bragg 1987, Steuter et al. 1990, Veblen and Lorenz 1991), southwestern oaks (McPherson et al. 1993, McClaran and McPherson 1995), and many pinyons and junipers (e.g., Johnsen 1962, Bragg and Hulbert 1976, Burkhardt and Tisdale 1976, Gruell 1983, West 1984, McPherson et al. 1988, Dick-Peddie 1993, Miller and Wigand 1994, Skovlin and Thomas 1995). In addition, density and cover of woody plants have increased within current and former savannas (Cooper 1960, Johnston 1963, Hastings and Turner 1965, Bragg and Hulbert 1976, Gehlbach 1981, Rogers 1982, Humphrey 1987, Bahre 1991, Covington et al. 1994, Miller and Wigand 1994, Archer 1995b, Arno et al. 1995, Skovlin and Thomas 1995).

The accounts of early Anglo travelers generally give the impression that vast areas of North America were characterized by dense stands of grass and widely scattered woody plants. For example, estimates of presettlement canopy cover of ponderosa pine

Figure 4.2 Several woody plants have increased in extent by spreading into former grasslands. This example illustrates encroachment of ponderosa pine in the Black Hills of South Dakota. An earlier cohort of ponderosa pine seedlings was killed by fire. (Photograph courtesy of Carolyn Hull Sieg.)

vary from 17% to 25% (Pearson 1923, White 1985, Covington and Sackett 1986). A presettlement ponderosa pine savanna was described by Beale (1858:49) as follows: "We came to a glorious forest of lofty pines, through which we have travelled ten miles. The country was beautifully undulating, and although we generally associate the idea of barrenness with the pine regions, it was not so in this instance; every foot being covered with the finest grass, and beautiful broad grassy vales extending in every direction. The forest was perfectly open and unencumbered with brush wood, so that the travelling was excellent." Similar descriptions can be found in the historical records of all North American savannas (e.g., Bartram 1791, Bryant 1848, Marcy 1849, Michler 1850, Bigelow 1856, Emory 1857, Evans 1990).

Increased extent and density of woody plants in North America have paralleled similar patterns throughout the world (e.g., Archer et al. 1988, Archer 1995b). Several mechanisms have been offered to explain the observed worldwide increase in woody plant abundance. The following sections evaluate these mechanisms within the context of North American savannas.

Tree Harvesting and Planting. Before about 1930, stands of woody plants were treated as nonrenewable resources. They were extracted as quickly as possible, with little consideration of how to replace them; in this respect, they were mined like precious minerals. As old-growth timber became increasingly rare on the continent, attitudes and policies began to incorporate the concept of sustainability, and trees were treated as renewable resources. Nonetheless, timber-management objectives continued to dominate longleaf pine and ponderosa pine savannas well after the 1930s. Thus, reforestation of these stands focused on the creation and maintenance of closed-canopy forests. The net result of intensive timber harvesting and subsequent reforestation has been replacement of savannas with other vegetation types: longleaf pine savannas have been largely replaced by loblolly and shortleaf pine forests, and ponderosa pine savannas have been replaced by ponderosa pine or mixed-conifer forests. Timber harvesting activities apparently did not contribute to conversion of former grasslands into savannas (i.e., they did not lead to increased distribution of North American savannas).

Longleaf pine was widely recognized as a commercially valuable species as early as the Revolutionary War; by the Civil War, half a billion board feet were cut annually (Wahlenberg 1946). The annual cut of longleaf pine continued to rise until 1907, when over 13 billion board feet were harvested (Wahlenberg 1946). Phloem (sap) usually was extracted from longleaf pine trees for a few years before they were harvested; the phloem was distilled to produce turpentine and rosin for the naval stores industry (Wahlenberg 1946).

Serious interest in reforestation with longleaf pine arose in the 1930s, after the Great Depression (Croker 1979 and Frost 1993 provide excellent historical accounts, which are summarized herein). The Civilian Conservation Corps planted millions of longleaf pine seedlings and fenced out feral hogs to prevent them from destroying young seedlings. Reforestation with longleaf pine continued through the 1950s, when about 8 million hectares were dominated by the species; this was about one-third of the presettlement extent of longleaf pine. By 1995, this figure was reduced to less than half by timber harvests and subsequent replacement with loblolly pine and shortleaf pine. These species are preferred by the timber industry because they are believed to grow more quickly than longleaf pine in commercial plantations. Because timber production has been the primary goal of reforestation efforts, tree densities are much higher in reforested (second-growth) stands than they were in presettlement stands, and reforested stands are characterized by different species than presettlement savannas. Thus, the overall results of timber harvesting activities were (1) greatly reduced extent of longleaf pine savannas and (2) conversion of most of the remaining savannas into closed-canopy forests of different species.

Initial harvest strategies in ponderosa pine stands demonstrated disregard for future supplies that paralleled earlier activities in longleaf pine savannas. However, large-scale timber harvesting of ponderosa pine did not occur until the late 1800s, when the railroad network was sufficiently developed to allow lumber to be exported from the region of harvest (Pearson 1950, Schubert 1974). The initial period of extensive tree removal (ca. 1880–1918) was followed immediately by a regionally heavy seed crop and exceptional tree recruitment in 1919 (e.g., Pearson 1923, Savage 1991). The resulting second-growth stands either have not been harvested or have been reforested after harvesting. Thus, the extent and abundance of ponderosa pine have increased, rather than decreased, since Anglo settlement (Covington et al. 1994). Timber production objectives contributed to second-growth stands with much higher densities than presettlement stands, which is consistent with changes in former longleaf pine savannas (e.g., Covington and Moore 1994a, 1994b).

With the exception of longleaf pine and ponderosa pine, dominant trees in North American savannas are not prized as commercial timber. Thus, with the exception of these two types of savannas, North American savannas were not subjected to widespread intensive tree removal after Anglo settlement. However, intensive harvests were conducted in local areas of all savannas.

Timber harvests associated with mining were particularly noteworthy and have been well documented (e.g., Billeb 1968,

Elliott 1973, Evans 1988, Bahre 1991). Stands of oak, mesquite, juniper, and pinyon were the primary sources of mine timbers and fuel for processing ore and cooking, so their removal was closely tied to the rise of mining industries in California (ca. 1849), Nevada (ca. 1859), and Arizona (ca. 1878). Harvesting extended 30–50 km from population centers and mines (Bahre 1991). However, locally intensive harvesting was restricted to areas immediately adjacent to mines and settlements; these areas accounted for less than 1% of the original extent of North American savannas. Thus, whereas activities associated with timber harvesting apparently contributed to increased woody plant density and cover in longleaf pine and ponderosa pine savannas, harvest of wood products from other savannas probably had negligible long-term, large-scale effects on woody plant abundance.

Atmospheric and Climatic Changes. One potential explanation for recent increases in woody plant abundance within grasslands and savannas is based on increased concentrations of atmospheric carbon dioxide ($[CO_2]_{atm}$) since the industrial revolution (e.g., Mayeux et al. 1991, Idso 1992, Idso and Kimball 1992, Johnson et al. 1993, Miller and Wigand 1994). It is argued that these increases in $[CO_2]_{atm}$ may have increased production of some plants (Polley et al. 1992). Increases in $[CO_2]_{atm}$ may confer a significant advantage to C_3 woody plants relative to C_4 grasses in terms of physiological activity, growth rates, and competitive ability. In accordance with this hypothesis, the encroachment of woody plants (with the C_3 photosynthetic pathway) into grasslands, particularly those dominated by C_4 grasses, has corresponded with a 27% increase in $[CO_2]_{atm}$ over the past 200 years.

The effects of rising $[CO_2]_{atm}$ on distribution and abundance of woody plants have been debated (Archer 1994, Archer et al. 1995). In particular, Archer et al. (1995) argue that changes in $[CO_2]_{atm}$ alone are not the proximate cause for observed shifts in woody plant distribution and abundance because (1) substantial increases in woody plant abundance in grasslands preceded a rise in $[CO_2]_{atm}$; (2) widespread replacement of C_3 grasses by C_3 shrubs has occurred, especially in temperate zones and cold deserts; and (3) C_4 species have quantum yields, photosynthetic rates, and water use efficiencies that are greater than those of C_3 species, even under current atmospheric $[CO_2]_{atm}$. Thus, there is no ecophysiological basis for a historic change in competitive interactions that favored C_3 over C_4 plants. These arguments, and the data that support them, provide convincing evidence that increased $[CO_2]_{atm}$ has not contributed significantly to historic increases in woody plant abundance, especially relative to the effects of livestock grazing and fire suppression (Bahre and Shelton 1993, Archer et al. 1995, Weltzin and McPherson 1995, Belsky 1996).

Considerable research on climate-vegetation relationships has been conducted at temporal scales of millennia (e.g., Wells 1983, Betancourt et al. 1990, Miller and Wigand 1994). The resulting climate-vegetation correlations have been described at scales of resolution too coarse to make meaningful recommendations to ecosystem managers. Furthermore, although these relationships provide some context for understanding large-scale and long-term vegetation changes, they shed little insight into rates, directions, or processes of vegetation change at contemporary spatial and temporal scales. Finally, these relationships are nonmechanistic. Thus, they will not be discussed herein.

In contrast to millennial-scale research on vegetation change, research on centennial temporal scales has been relatively scarce. Centennial-scale investigations are characterized by the same shortcomings as millennial-scale studies. An additional disadvantage of these studies is that they have tended to focus on shifts in the distribution of communities or biomes that may have resulted from climate changes (e.g., Idso and Quinn 1983, Sowell 1985, Neilson 1986, Idso 1992, Neilson et al. 1992, Neilson 1995). Yet all available evidence indicates that species respond individualistically (rather than as communities) to climate change (Davis 1989, Betancourt et al. 1990, Webb and Bartlein 1992, Rozema et al. 1993, Tausch et al. 1993). These studies present no convincing evidence of a directional change in climate that coincided with large-scale increases in woody plant abundance (e.g., Cooke and Reeves 1976, Bahre 1991, Bahre and Shelton 1993): savannas and grasslands were characterized by increased woody plant abundance before 1900 (Archer 1994, Archer et al. 1995), but global temperatures did not measurably increase until 1910 (Ghil and Vautgard 1991). Thus, large-scale increases in woody plant abundance preceded an increase in temperature by several decades.

There is little question that vegetation dynamics are correlated with short-term climatic events. For example, episodes of woody plant recruitment or mortality may be associated with anomalously wet or dry periods, respectively. These episodic events may have contributed to increased woody plant abundance within savannas (as discussed in chapter 3) and to expansion and contraction of savannas at their periphery (e.g., McPherson et al. 1988, Miller and Wigand 1994, McClaran and McPherson 1995). However, significant anthropogenically induced directional changes in climate have occurred only within the last few decades (Ghil and Vautgard 1991, Karl et al. 1993, Houghton et al. 1996). Therefore, it appears unlikely that directional climate change has contributed significantly to altered extent or composition of savannas within the last few centuries.

Livestock Grazing. Widespread increases in cover or density of woody plants coincident with the development of the livestock industry led many authors (reviewed by Archer 1994; see chapter 3) to conclude that the activities of grazing livestock have been the ultimate cause of increased woody plant dominance in former grasslands and savannas throughout the world. In North America, similar trends have been observed in ponderosa pine (Leopold 1924, Rummell 1951, Weaver 1951, Cooper 1960, Madany and West 1983, Steinauer and Bragg 1987, Baisan and Swetnam 1990, Savage and Swetnam 1990), pinyon-juniper (Ellison 1960, Johnsen 1962, Burkhardt and Tisdale 1969, 1976, Young and Evans 1981), and mesquite savannas (Buffington and Herbel 1965, Archer 1989).

It seems highly probable that livestock grazing contributed to (1) increased abundance of woody plants in current and former savannas and (2) expansion of savannas as woody plants established and persisted in former grasslands. In the former case, areal extent declined as savannas were replaced by closed-canopy woodlands or forests. In the latter case, savannas expanded at the expense of grasslands. (Mechanisms for these changes in physiognomy are described in chapters 2 and 3.)

In addition to directly affecting woody plant establishment, livestock grazing may facilitate woody plant persistence by contributing to reduced fire frequency (see chapter 3). The synergistic effects of livestock grazing and decreased fire frequency exemplify the strong interactions that may occur between land use and natural disturbance. These interactions help explain the increased distribution and abundance of woody plants in many former grasslands and savannas.

Fire Regimes. Fires were historically prevalent in all North American savannas. Ethnographical and historical accounts provide numerous descriptions of frequent, widespread fires (Stewart 1951, Lewis 1973, Croker 1979, Dobyns 1981, Wright and Bailey 1982, Pyne 1984, Bahre 1991, Hadley and Sheridan 1995). Dendrochronological evidence indicates that presettlement mean fire intervals were 7.4 years in Californian oak communities (McClaran and Bartolome 1989b), 2.8–4.3 years in midwestern oak (Guyette and Cutter 1991, Cutter and Guyette 1994), and 2–12 years in ponderosa pine (Weaver 1951, Cooper 1960, Baisan and Swetnam 1990, Swetnam 1990, Fulé and Covington 1995, Villanueva-Díaz and McPherson 1995, Swetnam and Baisan 1996). Similarly, mean fire intervals of 1–10 years are commonly reported in longleaf pine savannas (Wright and Bailey 1982, Ware et al. 1993, Glitzenstein et al. 1995 and references therein). A noteworthy, inexplicable exception to relatively frequent fires was reported on an isolated 150-hectare mesa in Utah, where the mean fire interval for a ponderosa pine savanna was 69 years (Madany and West 1980). Fires occurred every 4–7 years before 1881 on a nearby plateau.

Prehistoric fire frequencies are difficult to determine for mesquite and southwestern oak savannas because dominant woody species often burn completely or survive fires without scarring; in addition, the age of these plants is difficult to ascertain (Wright and Bailey 1982). However, several lines of indirect evidence, summarized by McPherson and Weltzin (1997), suggest that fires occurred about every 10 years in southwestern oak and mesquite savannas. This figure is consistent with dendrochronological assessments for Californian oak, midwestern oak, and ponderosa pine savannas. Dendrochronological evidence of historical fire frequency also is difficult to obtain for pinyon-juniper savannas, primarily because pinyon and juniper trees tend to either burn completely or survive fires without scarring (Jameson 1962, 1987, Burkhardt and Tisdale 1969, Young and Evans 1981). Indirect evidence indicates that fires occurred every 10–30 years in most pinyon-juniper savannas (Wright et al. 1979, Young and Evans 1981, Wright and Bailey 1982) and every 1–5 years in high-precipitation areas in savannas dominated by eastern redcedar (Bragg and Hulbert 1976, Wright and Bailey 1982, Steuter and McPherson 1995).

Considerable evidence suggests that widespread livestock grazing concomitant with Anglo settlement reduced fine-fuel biomass and therefore fire frequency (Wright and Bailey 1982, Archer and Smeins 1991, Archer 1994, Steuter and McPherson 1995). In fact, forest administrators encouraged overgrazing to reduce fire spread and enhance tree growth (Leopold 1924). Thus, the virtual absence of fires since Anglo settlement has been a conspicuous feature of most North American savannas. It seems likely that the interruption in recurrent fires contributed to increased abundance of woody plants within former and current savannas. In addition, fire exclusion probably contributed to encroachment of woody plants into former grasslands, thereby increasing the extent of savannas. Significant departures from these general patterns are associated with military reservations, where training exercises have contributed to the ignition and spread of frequent fires, and with longleaf pine savannas managed as quail plantations. In both cases, frequent fires have maintained relatively open savannas that have physiognomies similar to those described by early explorers.

In summary, significant changes have occurred in North American savannas, and different types of savannas have exhibited similar shifts in vegetation. Since Anglo settlement, some savannas have been cleared for agricultural crops or human settlements. Even more extensive changes in vegetation have involved shifts from native grasses to introduced grasses, or from savannas to closed-canopy woodlands or forests (fig. 4.3).

Despite the extensive literature that documents changes in community structure, there has been only partial agreement on

Figure 4.3 Paired photographs illustrate changes in vegetation during the last century. This site is located about 10 km west of Tombstone, Arizona. (A) Vegetation was grassland ca. 1890. (B) Vegetation was dominated by velvet mesquite and other woody plants ca. 1960; the scene was very similar in 1996. (Photographs, courtesy of Conrad Bahre, are from Hastings and Turner 1965 [pl. 53].)

the mechanisms responsible for the most widespread changes. Yet it is widely acknowledged that understanding mechanisms of vegetation change is central to interpreting and predicting responses of species and communities to future changes. Furthermore, it seems highly unlikely that a consensus will be reached in the debate about causal mechanisms of vegetation change. Two types of obstacles undermine the ability of retrospective studies to interpret and predict the response of contemporary or future communities to changes in climate, land use, or disturbance: (1) general constraints on the retrospective approach and (2) technical difficulties associated with specific methods. These obstacles are described below.

General Constraints on Retrospective Approaches

There are two primary, overarching limitations to using a retrospective approach for studying vegetation change. First, it is virtually impossible to accurately reconstruct the events and conditions that contributed to dramatic vegetation changes in the past. Second, even if this were possible, conditions responsible for historic or prehistoric species distributions are unlikely to be repeated in the future. Earth is entering an era unprecedented in terms of atmospheric gas concentrations and climatic conditions; thus, even a complete understanding of past climates and assemblages of organisms will not allow confident prediction of the future. This situation is exacerbated by species-specific response patterns that often are not linear or predictable, even within life-forms (Archer 1993). Finally, results of retrospective investigations do not elucidate mechanisms of vegetation changes (*sensu* Simberloff 1983, Campbell et al. 1991), because confounding between various factors precludes identification of mechanisms and also introduces the potential for spurious correlations.

Consider relatively recent large-scale changes in vegetation physiognomy that have occurred in North American savannas and in similar systems throughout the world. These dramatic changes have captivated the scientific community, but mechanisms of vegetation change remain unknown after more than three decades of detailed investigation (Archer 1989). For example, increased woody plant abundance in most grasslands and savannas has been attributed to changes in atmospheric or climatic conditions, reduced fire frequency, increased livestock grazing, or combinations of these factors (as reviewed by Archer 1994). Differing opinions about causes of vegetation change have contributed to acrimonious debate. For example, Bahre (1991:105), in a critique of work conducted by Hastings and Turner (1965), concluded that "probably more time has been spent on massaging the climatic change hypothesis than on any other factor of vegetation change, and yet it remains the least convincing." Such debate hardly seems beneficial for scientific advancement or appropriate management, yet it is a natural product of retrospective approaches.

Regardless of scientific progress toward consensus on mechanisms of past vegetation change, elucidation of these mechanisms would provide little or no predictive power to current and future management of savannas. Events that may have contributed to past vegetation change (e.g., cattle grazing, decreased fire frequency, specific timing of precipitation) may fail to produce similar responses today because of other, more profound changes in physical and biological environments over the last century. For example, plant communities now experience increased atmospheric concentrations of greenhouse gases (e.g., carbon dioxide, methane, nitrous oxides), increased abundance of woody perennial plants and introduced plants, and decreased abundance of some plant and animal species. Finally, there are no historic analogs for conditions that are now widespread. For example, a return to fire regimes characteristic of prehistoric savannas is unlikely to occur without major changes in land use because contemporary savannas are fragmented and are grazed by livestock. The descriptive nature of retrospective research coupled with the complex interactions underlying vegetation change make it extremely unlikely that a consensus will be reached in the debate about causal mechanisms of vegetation change. In many ways, reliance on retrospective techniques is analogous to driving a car by looking in the rearview mirror.

Specific Retrospective Techniques

In general, retrospective studies cannot be used to test hypotheses about vegetation change (see chapter 7), although they are useful for generating hypotheses. Further, retrospective techniques can be used to assess species-level changes in plant distribution only with dominant woody plants. A few studies of vegetation change are exceptional in their fine taxonomic resolution and spatial scale (e.g., Neilson and Wullstein 1983, Neilson 1986, Wondzell and Ludwig 1995). Nonetheless, these efforts are similar to other retrospective studies in that they are correlative and therefore cannot be used to distinguish between the many confounding factors associated with vegetation change. Additional limitations of specific types of retrospective studies—including historical accounts, dendrochronology, repeat photography, and analysis of organic carbon or biogenic opal in soil—are described below.

Historical accounts of vegetation change (e.g., land survey records, early maps, and notes of early travelers, surveyors, and military scouts) are usually anecdotal and imprecise and thus do not allow accurate determination of historic vegetation physiognomy or plant community composition. In addition, historical accounts are often contradictory and colored by fallacies (Bahre 1991).

Dendrochronology is limited to woody plants, usually trees, and is based on correlations between tree age and cross-sectional ring number. Dendrochronological assessments are used to describe dates of establishment, defoliation, or stem injury of individual woody plants. These assessments are then extrapolated to stands of trees in an attempt to describe periods of recruitment, mortality, rapid growth, or disturbance (Fritts 1976, Johnson and Gutsell 1994). However, if trees were once present but are currently absent, then reconstructions of stand age structure cannot be used to elucidate this important change. In addition, characteristics of dominant woody plants in Californian oak, southwestern oak, and mesquite savannas make them poorly suited for dendrochronological assessment: (1) current dendrochronological techniques usually are unsuitable for determination of stem age of these species (but see Flinn et al. 1994), and (2) because these species resprout after top removal, stem age does not necessarily represent individual plant age.

Repeat photography also has characteristics that seriously limit its usefulness for determining changes in plant distribution (Bahre 1991). Repeat photography at ground level has a limited and oblique field of view, and historic photographs usually portray anthropogenic manipulation of landscapes. Repeat photography from an aerial perspective is constrained by the date of the earliest photographs (ca. 1930 for North American savannas). Furthermore, extensive coverage of aerial photographs was not collected until after broad-scale vegetation changes had already occurred. In addition, extensive analyses of matched photographs in southwestern oak and mesquite savannas (e.g., Hastings and Turner 1965, Gehlbach 1981, Humphrey 1987, Bahre 1991) have fueled, rather than simplified, the controversy over alternative mechanisms of vegetation change (McPherson and Weltzin 1997).

Recently, analyses of stable carbon isotopes have been used to assess vegetation change in savannas (e.g., Steuter et al. 1990, Tieszen and Archer 1990, McPherson et al. 1993, McClaran and McPherson 1995). Stable isotope analysis relies on differential fractionation of carbon isotopes during photosynthesis. Nearly all woody plants possess the C_3 pathway of photosynthesis, whereas the dominant grasses in longleaf pine, southwestern oak, and mesquite savannas have the C_4 metabolic pathway. These two metabolic pathways ultimately affect the stable carbon isotope ratio ($^{13}C/^{12}C$) of living plant tissue, which is retained and incorporated into soil organic material after plant mortality and decomposition. Therefore, the stable carbon isotope ratio in the soil can be used as an indicator of previous vegetation on a site (fig. 4.4).

The isotopic composition of soil organic carbon does not accurately reflect past dynamics of C_3 and C_4 vegetation if (1) the isotopic composition of the surface soil differs from that of the overlying vegetation; (2) soil depth is not an appropriate surrogate for time (i.e., if relatively new carbon is transported beneath older soil carbon via soil mixing which may be caused by burrowing animals, freezing and thawing cycles, or alluvial processes); (3) deep-rooted C_3 plants (e.g., shrubs, trees) deposit soil carbon

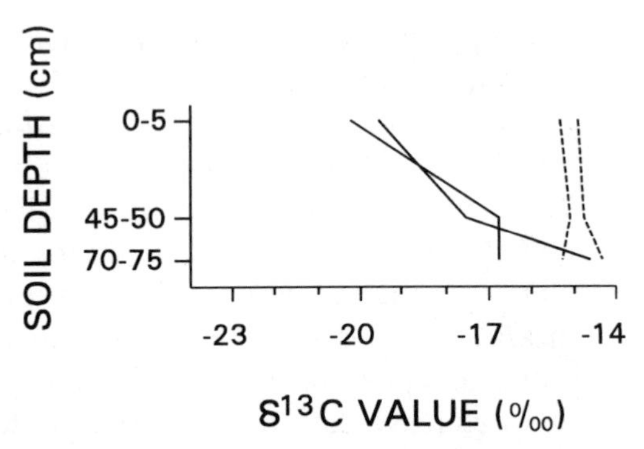

Figure 4.4 Stable carbon isotope composition ($\delta^{13}C$ value) of soil organic matter at different soil depths at the oak savanna–grassland ecotone in southern Arizona. Each line represents a single soil core. Solid lines represent cores collected beneath trees at the savanna-grassland ecotone, and dashed lines indicate cores collected in the adjacent grassland. Strongly negative $\delta^{13}C$ values directly beneath trees suggests that they are recent occupants of former grassland. (Modified from McClaran and McPherson 1995.)

beneath C_4 plants; or (4) current or former dominant grasses possess the C_3 photosynthetic pathway. Because these conditions often occur within savannas, stable isotope analysis is not appropriate for studying within-savanna dynamics (McClaran and McPherson 1995). However, analysis of stable carbon isotopes is useful for identifying shifts in the periphery of savannas dominated by C_4 grasses and C_3 woody plants.

Vegetation changes may be inferred by assessing biogenic opal (i.e., plant microfossils or opal "phytoliths") in soils (e.g., Kalisz and Stone 1984), and the technique is conceptually similar to stable isotope analysis. Grasses produce more biogenic opal than do woody plants, and the opal from grasses is morphologically distinctive from the opal of woody plants (Witty and Knox 1964, Kalisz and Stone 1984). Biogenic opal is composed of silica dioxide, which is very resistant to decomposition. Thus, the abundance and type of opal in the soil can be used to indicate previous vegetation on a site. Biogenic opal can be used to distinguish between some members of the grass family and has therefore been used to study conversions from perennial to annual grasslands (e.g., Bartolome et al. 1986). Few studies of vegetation change in North American savannas have used biogenic opal. However, the limitations of biogenic opal analysis are similar to those of stable isotope analysis: both techniques rely on chemical or morphological differences between plant taxa (especially grasses and woody plants) and make similar assumptions about deposition in the soil.

The concurrent use of several different retrospective techniques may facilitate appropriate interpretation of previous changes in vegetation physiognomy. However, different retrospective techniques may also generate conflicting interpretations of the same phenomena, as illustrated by the following example (McPherson and Weltzin 1997).

Reports of past changes in the oak savanna–semidesert grassland boundary are varied. Paleoecological data suggest that oak savannas have shifted upslope in concert with warmer and drier conditions since the Pleistocene. This interpretation is consistent with upslope movement of most woody species in the last 40,000 years, as determined by paleoecological research (Betancourt et al. 1990). On a more contemporary temporal scale, Hastings and Turner's (1965) use of repeat photography at ground level led them to conclude that the oak savanna–semidesert grassland boundary moved upslope during the last century. In contrast, research based on stable carbon isotope technology and radiocarbon dating indicated that oaks at the savanna-grassland boundary had encroached into former grasslands within the last 1,000 years, which implied that oak savannas had shifted downslope into semidesert grasslands (McPherson et al. 1993, McClaran and McPherson 1995). The latter finding matched Leopold's (1924) interpretation of downslope movement of oaks, based on observations of progressively smaller trees from the savanna into the grassland. Finally, Bahre (1991) examined surveyor's records as well as repeat photography from the ground and air and concluded that the distribution of oak savannas had been stable since the 1870s. Thus, boundaries between oak savannas and adjacent semidesert grasslands have been variously reported as shifting upslope, remaining static, or shifting downslope. Although these differences may be attributable in part to variation in temporal and spatial scales, they are largely the result of different interpretations.

This example suggests that the disadvantages associated with retrospective techniques cannot be overcome simply with the use of a combination of methods. Thus, not only are retrospective approaches constrained by fundamental conceptual and philosophical limitations, they are hampered by several technical weaknesses. Several of the technical obstacles formerly associated with specific retrospective techniques have been removed by significant technological advances; however, the major conceptual and philosophical limitations can be overcome only by traveling back in time.

Summary

Past changes in North American savannas have been widely documented. These studies indicate that many savannas have been replaced by urban areas, agricultural fields, and closed-canopy forests. Other savannas have arisen as woody plants have spread into former grasslands. The understory layer of most savannas has been altered by widespread establishment of introduced plants.

Numerous mechanisms have been proposed to account for large-scale changes in the structure of North American savannas. However, the relative importance of various mechanisms is difficult to evaluate with retrospective studies because of general and specific limitations associated with retrospective techniques. Furthermore, retrospective approaches are poorly suited for the assessment of contemporary conditions and the prediction of future ecological changes. Rather, judicious interpretation of results from relevant experiments should form the basis for predicting future changes in savanna vegetation. This approach is pursued in the following chapter.

5

Expected Future Changes

"It's a poor sort of memory that only works backwards," the Queen remarked.

Lewis Carroll, Through the Looking Glass, and What Alice Found There

It is unlikely that we will ever be able to accurately reconstruct the events and conditions that contributed to dramatic vegetation changes after Anglo settlement. Several authors have reviewed the role and importance of various factors responsible for vegetation changes, but they have failed to reach a consensus (see chapter 4). Even if a consensus could be reached, presettlement events and conditions are unlikely to occur again because of anthropogenic changes such as urbanization, changes in climate and concentrations of atmospheric gases, altered patterns of herbivory, introduction of plant species, altered fire regimes, and timber removal and replacement. Therefore, prediction of vegetation dynamics and development of management strategies must include consideration of present and future environments. This chapter describes attributes likely to change in North American savannas in the near future, then assesses the impacts of these factors on interactions between woody plants and grasses and discusses subsequent ramifications for changes in savanna physiognomy.

Changes in some relevant phenomena can be confidently predicted (e.g., increased atmospheric CO_2 concentrations), whereas changes in others are less likely to occur without a substantial shift in priorities of resource managers (e.g., rates of livestock grazing). Furthermore, the resulting changes in savanna physiognomy vary from highly probable (e.g., urbanization resulting in loss of savanna) to unknown (e.g., effects of altered frequency of precipitation). Thus, predicting the future extent and structure of savannas is an exercise that ranges from elucidating the obvious to (barely) educated guessing. Nonetheless, development of relevant policy guidelines and appropriate management strategies is dependent on knowledge of future conditions.

Regardless of cause, changes in vegetation physiognomy have broad implications for management and land use. Life-form changes in plant communities affect virtually all resources, including soil, water, wildlife, livestock, fuelwood, and recreation. In addition, changes from herbaceous- to woody plant–dominated communities constitute a potentially important global climate feedback affecting carbon sequestration, nonmethane hydrocarbon emissions, and biophysical land surface–atmosphere interactions (e.g., albedo, evapotranspiration, surface roughness, boundary layer dynamics) (Archer 1994).

Science has an obligation to provide relevant information to policy makers and managers while acknowledging the limits of this information. This chapter represents an attempt to address these issues. An exhaustive discussion of future changes and likely effects is not attempted. Rather, this chapter draws upon existing knowledge (described in chapters 1–4) to provide a conceptual framework and brief discussion of potentially important phenomena.

Urbanization

Increased urbanization of savannas and other natural ecosystems undoubtedly will continue. Urbanization produces the most obvious and least reversible changes in the structure and function of savannas: locally, savannas are eliminated and replaced with an urban community (fig. 5.1). Regionally, impacts may include fragmentation of natural communities, increased demand for recreational activities and areas, reduction of fire frequency, altered movements and grazing patterns of native and domestic animals, increased conflict between "new" urban and "traditional" rural inhabitants, and reduction of groundwater levels with concomitant changes in surface water regimes.

The increasing human population in North America serves as a constant source of pressure on natural communities and their

Figure 5.1 Urbanization eliminates savannas at the local level. Regionally, impacts may include fragmented natural communities, increased demand for recreational activities and areas, reduced fire frequency, altered movements and grazing patterns of native and domestic animals, increased conflict between "new" urban and "traditional" rural inhabitants, and reduced groundwater levels with concomitant changes in surface water regimes. (Photograph courtesy of Lynn Huntsinger.)

products (Ehrlich 1995). The rate of population increase in the United States (0.9%) is among the lowest in the world, but it still accounts for nearly 2 million people each year (excluding immigration) (Williams 1991, DESIPA 1993). Legal and illegal immigration add at least 0.3 million people to this annual increase. Although the rate of population increase in Mexico (2.2%) is more than double that of the United States, the difference between birth and death rates is about equal for the two countries. Thus, differences in total population and rate of population increase between the United States and Mexico offset one another to produce similar annual increases in the human population of the two countries. In contrast to the United States, Mexico loses about 150,000 people annually to emigration (Williams 1991, DESIPA 1993).

Demographic shifts apply additional pressure on North American savannas, as people seek a higher quality of life than can be found in large cities. For example, the most rapidly growing urban areas in California (FRRAP 1988) and Arizona (McClaran and McPherson 1998) are located within or adjacent to Californian oak and southwestern oak savannas, respectively. Relocation of United States citizens from the north to the south and west (Williams 1991) is an additional demographic factor that affects savannas: nearly all North American savannas are associated with relatively mild climates in southern or coastal regions (chapter 1).

Because of differences in land ownership, increased urbanization will not occur equitably among all savannas. Southwestern oak, pinyon-juniper, and ponderosa pine savannas are located largely within lands managed by government agencies (primarily national forests in the United States and cooperatives in Mexico) and are therefore improbable areas for urban development. In contrast, other major North American savannas are located primarily on private land. Urban development has been the greatest single source of loss of Californian oak savannas, which are primarily privately owned. This trend is expected to continue in the foreseeable future (Bolsinger 1988). Midwestern oak savannas near large urban centers in Texas are experiencing similar patterns of development. When population pressures in mesquite and longleaf pine savannas resemble those in Californian oak savannas, similar trends in urban development can be expected there.

Changes in Atmospheric Gas Concentrations

Human activities have contributed to increased concentrations of greenhouse gases, including carbon dioxide, methane, and nitrous oxides (Neftel et al. 1985, Trabalka et al. 1986). Additionally, rapid increases in greenhouse gas concentrations are likely within the next 50 years (Houghton et al. 1990, 1992, 1996). In addition to affecting the hydroclimatology of the earth, increased concentrations of these gases may directly influence plant interactions within North American savannas. Effects of carbon dioxide have been especially well studied with respect to vegetation dynamics (e.g., Koch and Mooney 1996 and references therein). Effects of other atmospheric gases have not been documented adequately to permit the development of reasonable estimates of their impacts on North American savannas.

Woody plants and grasses may respond differently to changes in atmospheric carbon dioxide concentrations ($[CO_2]_{atm}$). Differential response may be particularly apparent in savannas where woody plants and grasses have differing photosynthetic metabolisms. For example, southwestern oak, mesquite, and southeastern pine savannas are codominated by woody plants that possess the C_3 pathway of photosynthesis and grasses that possess the C_4 photosynthetic pathway. When exposed to elevated $[CO_2]_{atm}$, C_3 plants exhibit greater increases in growth and photosynthesis than do C_4 plants grown under the same conditions (as reviewed by Bazzaz 1990 and Patterson and Flint 1990). When grown with elevated $[CO_2]_{atm}$, water use efficiency (WUE) of C_3 plants generally is increased more than WUE of C_4 plants (Polley et al. 1992, Polley et al. 1993). Thus, at constant temperatures and elevated $[CO_2]_{atm}$, physiological changes favor C_3 plants over their C_4 counterparts, at least in controlled environments (Wray and Strain 1987, Bazzaz 1990, Patterson and Flint 1990).

Changes in plant physiology that result from shifts in $[CO_2]_{atm}$ may produce changes in the relative abundance or spatial distribution of plant species (Davis 1989, Bazzaz 1990, Long and Hutchin 1991, Neilson 1993, Neilson and Marks 1994). For example, it has been hypothesized that increases in $[CO_2]_{atm}$ may contribute to dramatic increases in extent and abundance of C_3 annual grasses such as downy brome (Mayeux et al. 1994). Similarly, growth and establishment of C_3 shrubs may increase in C_4-dominated grasslands and savannas of the southwestern United States (Mayeux et al. 1991, Idso 1992, Johnson et al. 1993, Polley et al. 1994). Further, CO_2-induced increases in WUE or fine root biomass of C_3 woody plants (e.g., Norby et al. 1986, Idso and Kimball 1992) suggest that these plants may be able to expand their distribution into ecosystems where water is otherwise a limiting factor (Long and Hutchin 1991, *sensu* Mellilo et al. 1993). This premise is supported by correlative evidence that higher WUE is positively correlated with survival of plants in dry habitats (Ehleringer and Cooper 1988). A simple WUE model developed by Idso and Quinn (1983) suggested that a doubling of $[CO_2]_{atm}$ would cause oak woodlands in the southwestern United States to shift downslope and displace extensive regions of semidesert grassland. However, their hypothesis has not been tested in southwestern oak savannas or any other system. In addition, considerable caution is warranted when using the response of individual plants to two levels of $[CO_2]_{atm}$ (ambient, doubled) as a basis for predicting the response of communities to incremental changes in $[CO_2]_{atm}$: plant response to $[CO_2]_{atm}$ differs between individual plants and communities, and the response varies with the magnitude and rate of increase of $[CO_2]_{atm}$ (Ackerly and Bazzaz 1995).

Climate Change

Anthropogenically induced changes in global climate have occurred, are occurring, and will continue to occur at increasing rates in the future. Climate change will affect all North American savannas and may mask or interact with other factors. In fact, climate change may replace traditional disturbances (e.g., livestock grazing, fire) as an important regulator of vegetation change (Weltzin and McPherson 1995).

It should be recognized that although there is a scientific consensus about the occurrence of anthropogenically induced changes in climate, the potential temporal and spatial magnitude of many of these changes is debated (e.g., Lindzen 1993). In addition, these changes may not be directional. For example, changes in climate may result in increased frequency of extreme climatic events (Wigley 1985, Katz and Brown 1992). Thus, the following phenomena, which are known to significantly affect recruitment or mortality of savanna plants, are likely to increase in frequency:

freezing temperatures during late spring or early autumn, unusually high temperatures, tropical storms, and hurricanes. High winds and saturated soils associated with the latter phenomenon are a major source of longleaf pine mortality (Gresham et al. 1991). Also, climate may become more serially correlated, which will result in an increased probability of sequences of warm or cold (wet or dry) years and greater overall climate variability (Cohen and Pastor 1991). These cycles may contribute to "pulsed" recruitment (during wet periods) or mortality of woody plants (during dry periods), as discussed in chapter 3. Because savannas represent a mixture of life-forms particularly sensitive to environmental changes, virtually any change in climate is likely to affect savanna vegetation at the individual, population, or community level, with subsequent ramifications for ecosystem structure and function. For example, a several-degree temperature increase in Californian oak savannas is predicted to cause blue oak savannas to virtually disappear but is also expected to produce an expansion in the extent of the less common interior live oak savannas (Evett 1994).

Recent reviews have discussed changes in vegetation likely to result from near-future changes in climate (e.g., Weltzin and McPherson 1995, McPherson and Weltzin 1997). These reviews are summarized below within the context of North American savannas and are extended to incorporate North American savannas excluded from earlier reviews. This exercise should be interpreted with considerable caution: empirical research on effects of climate change on vegetation has rarely included North American savannas and has been largely *post hoc* and descriptive.

Anthropogenic activities have contributed to increased global temperatures (Jones 1993, Karl et al. 1993, Nasrallah and Balling 1993a, 1993b), especially since the beginning of the industrial revolution. Changes have not been consistent, even in North America: mean temperatures have increased in the southwestern United States and have decreased in the southeastern United States since 1910 (Dettinger et al. 1995). Increases in $[CO_2]_{atm}$ have produced a 0.4°C increase in mean global surface temperature since 1910 (Ghil and Vautgard 1991) and are expected to produce a net increase between 1.5°C and 4.5°C, with a "best guess" of 2.5°C (Mitchell et al. 1990, Houghton et al. 1992, 1996). This rapid increase in mean temperature has been predicted since the late 1800s, when crude computational tools were used to generate predictions nearly identical to modern forecasts (Arrhenius 1896). Increased mean temperatures are expected to be associated with increased climatic variability at global and regional scales within the next 50 years (Houghton et al. 1990, 1992, 1996, Jones 1993, Karl et al. 1993, Nasrallah and Balling 1993a, 1993b).

CO_2-induced changes in the amount and distribution of precipitation have also been predicted (Houghton et al. 1990, 1992, 1996). These changes are expected to increase the severity and frequency of droughts in North America, among other effects (Manabe and Wetherald 1986, Balling et al. 1992, Idso and Balling 1992). However, there is considerable uncertainty about the direction and magnitude of future shifts in precipitation distribution, especially on a regional basis.

Photosynthesis and growth of C_4 plants apparently are limited more by cool temperatures than by $[CO_2]_{atm}$ (Long 1983, Potvin and Strain 1985). Therefore, in contrast to the expected beneficial effects of increased $[CO_2]_{atm}$ on C_3 plants described above, global increases in temperature may enhance the competitive ability of C_4 plants (including dominant grasses in southern savannas) relative to C_3 plants (e.g., shrubs and trees), especially where soil moisture (Neilson 1993) or temperatures (Esser 1992) are currently limiting. This could result in regional increases in grasslands at the expense of savannas (*sensu* Long and Hutchin 1991) or decreased density of woody plants in forests and woodlands. The latter phenomenon could transform closed-canopy stands of southwestern oak, mesquite, and southeastern pines into savannas.

Perhaps more important than the individual effects of increased $[CO_2]_{atm}$ and temperature, however, is the potential interactive effect of these conditions on photosynthetic productivity and ecosystem-level processes (Long 1991). Unfortunately, the relatively few studies of interactive effects of temperature and $[CO_2]_{atm}$ on vegetation (Bazzaz 1990, Farrar and Williams 1991) have produced conflicting, poorly understood results. For example, photosynthesis and growth of plants in elevated $[CO_2]_{atm}$ may be stimulated by increases in temperature (Idso et al. 1993). Alternatively, temperature may have little or no effect on CO_2-

enriched plant growth (Jones et al. 1985, Tissue and Oechel 1987). Nonetheless, *a posteriori* analysis of vegetation response to $[CO_2]_{atm}$ as mediated by atmospheric temperatures suggests that relative effects of $[CO_2]_{atm}$ increase as temperature increases (Drake and Leadley 1991, Idso et al. 1993), perhaps due to upward shifts in optimal temperature for photosynthesis with increasing $[CO_2]_{atm}$ (Pearcy and Bjorkman 1983). For example, Mooney et al. (1991) predict that $[CO_2]_{atm}$ will increase ecosystem productivity where daytime temperatures are above 30°C, as in most North American savannas.

Regional shifts in precipitation seasonality have been linked to global warming and cooling trends (Neilson and Wullstein 1983, Neilson 1986), and there is every reason to believe that the seasonal distribution of precipitation will change in the near future. Alterations in the hydrologic cycle probably have greater importance to society than do increases in temperature; although global temperature and precipitation patterns are related, changes in the amount and distribution of precipitation have direct impacts on many human activities (Groisman and Legates 1995). Furthermore, minor changes in precipitation pattern, particularly at ecotones (Neilson 1987, 1993), may cause major shifts in the distribution and abundance of plants (*sensu* Stephenson 1990). Nonetheless, there are at least three significant barriers to predicting the effects of these changes on North American savannas: (1) changes likely will vary among regions; (2) a specific change in precipitation seasonality is unlikely to produce consistent responses in different savannas because of individualistic species-level responses; and (3) there is no consensus on the direction or magnitude of likely shifts in precipitation (Houghton et al. 1992, 1996). Thus, although global and regional shifts in precipitation seasonality are likely to occur, and despite the fact that these shifts will undoubtedly affect savannas at coarse (e.g., ratios of woody plants to grasses) and fine (e.g., species-specific) levels of resolution, insufficient data exist to permit evaluation and quantification of these concepts.

Two important conclusions are evident from previous research on climate change and vegetation. First, interactive effects of changing abiotic conditions may be more important than simple effects in terms of plant response and subsequent changes in vegetation physiognomy. For example, the interaction between increased $[CO_2]_{atm}$ and temperature may be more important than the effects of either factor by itself. This conclusion underscores a crucial point regarding our ability to predict and explain the response of vegetation to climate change: there are insufficient data to permit an evaluation of the response of ecosystems to reasonable scenarios of global change. Despite the strong probability that multiple factors will be changing simultaneously in the near future, no field studies have evaluated the importance of interactions between climatic factors on relationships between woody plants and grasses. Second, previous research on climate change and vegetation indicates that impacts of $[CO_2]_{atm}$ enrichment may be mediated by other environmental constraints (e.g., temperature or nutrient limitations). Thus, because of environmental constraints, "real-world" ecosystems likely will have muted responses compared to studies conducted in small pots or chambers. Unfortunately, studies designed to evaluate the relevance of mesocosm experiments to field conditions are lacking (Bazzaz 1990, Rozema et al. 1993).

In summary, changes in climate that may occur in North American savannas include increased surface temperatures, changes in the seasonal distribution of precipitation, more frequent climatic extremes, and generally greater climatic variability. Interactions between abiotic factors may be more important than effects of any single factor with respect to vegetation response. Nonetheless, it is widely hypothesized that increases in $[CO_2]_{atm}$ and winter precipitation should favor woody plant establishment and growth at the expense of grasses, particularly in savannas dominated by C_4 grasses (e.g., Neilson 1995 and references therein). Alternatively, increases in temperature (associated with increased $[CO_2]_{atm}$) and summer precipitation may favor C_4 grasses at the expense of C_3 woody plants.

Herbivory

Morphological and physiological characteristics of plants exert considerable control over plant response to herbivory. Furthermore, these morphological and physiological characteristics affect interactions between contrasting life forms (e.g., woody plants, grasses), and these interactions may be mediated by anthropogenic affects. Thus, altered patterns of herbivory may have important implications for the relative proportion of woody plants and grasses in North American savannas. Two changes in herbivory are predicted to occur: (1) increased consumption of C_3 plants by native herbivores, and (2) decreased levels of livestock grazing.

A critical constraint on native herbivores is the ability to acquire nitrogen (Mattson 1980, Slansky and Rodriguez 1987, White 1993). C_3 plants grown in controlled environments at enriched levels of $[CO_2]_{atm}$ produce increased carbon:nitrogen (C:N) ratios relative to plants grown at ambient $[CO_2]_{atm}$ (e.g., Lincoln et al. 1986, Williams et al. 1986, Osbrink et al. 1987, Fajer et al. 1989, 1991); C_4 plants generally do not exhibit this response (Wong 1979, Curtis et al. 1989a). Thus, herbivores must increase consumption of plant material with high C:N ratios to obtain adequate nitrogen (Lincoln et al. 1986, Osbrink et al. 1987, Lincoln and Couvet 1989, Fajer et al. 1989, 1991, Johnson and Lincoln 1991, Lincoln et al. 1993). Consequently, competitive advantages that accrue to C_3 plants from increased $[CO_2]_{atm}$ may be negated by increased herbivory. Alternatively, elevated $[CO_2]_{atm}$ may contribute to decreased populations of herbivores as a result of decreased food quality (Oechel and Strain 1985), although this response is not widely documented (Lincoln 1993) and seems unlikely. The net result of increased $[CO_2]_{atm}$ should be increased herbivory on C_3 herbs and woody plants. This outcome may be negated if the impact of generalist herbivores is very large: these generalists may switch grazing preference to species that demonstrate little or no increase in C:N ratio.

In contrast to anticipated increases in herbivory from native animals, livestock grazing is likely to decrease in the near future. Increasing residential development and a rapidly expanding market for ranchettes is reducing the stability of the livestock grazing industry (McClaran et al. 1992). Urbanization produces higher property values and property taxes; it also increases land-use conflicts. Increased taxes can preclude profitability in an existing livestock operation and, combined with the lure of profits from land sales and the personal frustration generated by land-use conflicts, may result in a positive feedback on the rate of residential development (McClaran et al. 1992, McClaran and McPherson 1998). Although some tax relief is available through differential property taxation for agricultural land in some states (McClaran et al. 1992), livestock grazing is expected to decline in the near future. The net result should be increased herb abundance, with a wide variety of consequences (see chapter 2).

Introduction of Plant Species

Several factors could combine to facilitate encroachment of introduced species into savannas. Savannas are experiencing higher levels of human use than at any time in the past. Further, use of savannas is trending from extensive (e.g., livestock grazing, dispersed recreation) to intensive (e.g., urbanization, off-road vehicle use). Disturbances associated with increased intensity of use (e.g., roads, trails, urban areas) create conditions amenable to establishment of introduced species. Direct dispersal, either intentional (e.g., horticulture, landscape architecture) or accidental (e.g., seed dispersal in vehicles or clothing), represents another source of increases in populations of introduced plants.

In addition to increased disturbances and dispersal of seeds, both associated with anthropogenic activities, altered climates may enable species to occupy areas previously unsuitable for establishment or growth. For example, Lehmann lovegrass, a native of South Africa, is dominant throughout the semidesert grassland in the southwestern United States (Schmutz et al. 1991, McClaran and Van Devender 1995) but is rare or absent in adjacent oak savannas. Oak savannas are higher and slightly cooler than semidesert grassland. Lehmann lovegrass appears to be cold-limited at high elevations (Cox et al. 1988), which implies that even a small increase in regional temperature will allow the species to establish in southwestern oak savannas. Similarly, downy brome, a native of Eurasia, may continue to spread throughout the western United States as a result of increased $[CO_2]_{atm}$ (Smith et al. 1987, Mayeux et al. 1994).

Altered Fire Regimes

Ongoing and impending changes in climate or land use may affect fire frequency and extent. For example, fragmentation associated with increased urbanization—coupled with efficient, aggressive fire suppression—has reduced the frequency of large fires in the United States. In contrast, lower rates of fragmentation and more extensive land management practices have contributed to the maintenance of historic fire regimes through the present time in neighboring Mexico (Fulé and Covington 1995, Swetnam and Baisan 1995, Villanueva-Díaz and McPherson 1995). As the Mexican transportation infrastructure approaches that of the United States, however, fire frequency and extent likely will decline.

Changes in plant community structure also may alter fire regimes, as they have in the past (*sensu* Clark 1990). For example, increased woody plant density reduces herbaceous biomass and subsequently reduces fire frequency because of decreased accumulation of fine fuel (McPherson 1995). Conversely, changes in precipitation regimes may increase fine-fuel loading and thus increase fire frequency, intensity, and extent (Rogers and Vint 1987, Swetnam and Betancourt 1990, Billings 1994, MacCleery 1995). Increases in global or regional surface temperatures may either increase fire frequency because hotter, drier conditions cure fuel more quickly, or decrease fire frequency because of decreased fine-fuel production associated with hotter, drier conditions. Finally, increased biomass associated with encroachment of introduced grasses may produce a positive feedback on fire frequency and extent: introduced grasses often exhibit higher production and have more continuously distributed biomass than native grasses, which leads to an increased frequency of fires, thereby accelerating the encroachment of introduced grasses (Pickford 1932, Wright and Klemmedson 1965, Young et al. 1976, Mayeux et al. 1994, McPherson 1995). Obviously, future fire regimes are difficult to predict, in part because of the paucity of knowledge about future climate change and also because of the interactive effects of climate change, biological factors, and activities related to management and politics.

Timber Removal and Replacement

Increases in woody plant density are expected to continue in ponderosa pine and longleaf pine savannas, which will facilitate conversion of these savannas to closed-canopy forests. Such conversions have occurred within the last century (see chapter 4), and increasing demand for timber in the future may fuel intentional increases in tree density (hence, conversion of savannas to forests). However, low economic costs associated with burning fossil fuels, combined with the remoteness of most savannas in western North America, will tend to maintain low harvest rates of woody plants from most savannas.

Awareness of the presence of rare species (e.g., spotted owl, red-cockaded woodpecker) tends to decrease timber removal from public lands. The subsequent increase in timber prices encourages increased timber removal from private lands. Thus, the gap between low harvesting rates on savannas under federal jurisdiction and high harvesting rates on private land is likely to widen as the controversy over rare species intensifies (Adams et al. 1996).

Savanna restoration programs, which initially focus on establishment of woody plants, have been developed only in savannas dominated by private ownership (Californian oak, midwestern oak, and longleaf pine savannas). Emergence of these programs largely coincided with the expansion of conservation biology as a scientific discipline circa 1980. Restoration efforts in midwestern oak savannas appear to be succeeding because of intensive management by volunteer-assisted nonprofit organizations (Fralish et al. 1994); restoration of longleaf pine savannas is feasible, but success to date has been limited by a lack of economic incentives (Landers et al. 1995). Nonetheless, the area occupied by longleaf pine and midwestern oak savannas may stabilize or increase as a result of recent interest in restoration and preservation by the private sector (e.g., Noss 1989, Standiford 1991, Anderson et al. 1994, Landers et al. 1995). Urbanization and economic development associated with the rapidly increasing population have undermined large-scale restoration efforts in Californian oak savannas (Standiford 1991); restoration and preservation efforts must increase substantially if they are to keep pace with rates of savanna loss.

Timber production in pinyon-juniper, southwestern oak, and mesquite savannas is insufficient to warrant intensive management for traditional forest products, so effects of timber harvesting in these savannas are minor (but see Meyer and Felker 1990 for a discussion of specialized wood products associated with mesquite savannas). In addition, pinyon-juniper, southwestern oak, and ponderosa pine savannas are dominated by federal ownership, and mesquite savannas are sufficiently widespread and protected from urbanization; therefore, current and expected loss rates of these savannas are low and restoration is not a widespread concern.

Summary

There is every reason to believe that the structure and function of North American savannas will change significantly in the near future as a result of alterations in atmospheric gas concentrations, global or regional climate, human demographic patterns, land uses, management strategies, and interactions between these factors. Further, the distribution of savannas is likely to change as savannas replace grasslands or forests. Increased woody plant abundance may create savannas from existing grasslands; alternatively, high rates of woody plant mortality in existing forests—produced by increased herbivory by native herbivores or increased drought stress associated with changes in climate—may facilitate a transition to savannas. Additionally, changes in woody plant abundance may transform some existing savannas into grasslands or closed-canopy woodlands.

Unfortunately, few changes in savanna distribution and physiognomy will be simple, predictable, or consistent among different savanna types. Considering the myriad factors likely to change in the near future, development of reliable predictions of vegetation change will require considerably more research. A short list of critical experiments necessary to provide the requisite knowledge is outlined in chapter 7, and suggestions for management in light of current knowledge are provided in chapter 6.

6

Applying Ecological Knowledge

Give us the tools, and we will finish the job.
Sir Winston Churchill, radio broadcast, 1941

There is considerable uncertainty about mechanisms underlying vegetation change in most North American savannas. Additional uncertainty surrounds future responses to changing atmospheric gas concentrations, climates, disturbance regimes, and land uses. Although the depth and breadth of scientific ignorance about these systems combine to offer a bright future for inquisitive ecologists, they also severely constrain the ability to accurately predict the response of individual sites to management actions. Nonetheless, the demands of an increasingly large and diverse society necessitate appropriate, cost-effective management.

Interestingly, some scientists believe that there is insufficient knowledge to support ecologically based recommendations about the management of natural resources, whereas others believe that ecologists are uniquely qualified to make these recommendations. Of course, decisions about natural resources must be made—even the absence of action is a management option—so it seems appropriate to apply relevant ecological knowledge to these decisions. However, ecologists generally have no expertise in the political, sociological, or managerial aspects of resource management, and they are rarely affected directly by decisions about land management. Thus, ecologists are not necessarily held accountable for their recommendations, nor are they ultimately responsible for being stewards of the land. Conversely, managers are ultimately accountable for their actions and are responsible for land stewardship, so they should exploit relevant ecological information as one component of the decision-making process.

Clearly stated goals and objectives will facilitate management and allow selection of appropriate tools for accomplishing these goals and objectives. Conversely, selection of goals or objectives that are poorly defined or quantified may impede management. For example, use of the term "ecosystem health" implies that there is an optimal state associated with an ecosystem and that any other state is abnormal; however, the optimal state of an ecosystem must be defined, and clear and quantifiable objectives must be developed to achieve that state. Similarly, "ecosystem integrity" (Wicklum and Davies 1995) and "sustainability" (Lélé and Norgaard 1996) are not objective, quantifiable properties. The use of terms such as "health," "integrity," and "sustainability" as descriptors of ecosystems implies that managers or scientists can identify the state that is optimum for the ecosystem (rather than optimum for production of specific resources) and that the preservation of this state is scientifically justifiable. These terms are not supported by empirical evidence or ecological theory, and they should be abandoned in favor of other, more explicit, descriptors (Wicklum and Davies 1995). Appropriate goals and objectives should be identified on a site-specific basis and linked to ecosystem structures or functions that can be defined and quantified.

For two primary reasons, this chapter does not provide explicit recommendations for resource managers—it is not a "how-to" guide for management. First, management decisions must be temporally, spatially, and objective-specific. Thus, because this book is focused on broad-scale spatial and temporal factors such as the responses of savannas to changes in disturbance regimes, it should not be used to justify site-specific management decisions. Rather, management decisions should be couched within this temporally and spatially broad discussion and should be made by the managers most familiar with individual systems. Second, specific management activities, though based on scientific knowledge, are conducted within the context of relevant social, economic, and political issues (*sensu* Brown and MacLeod 1996). These issues and concerns are beyond the scope of this book, which instead concentrates on scientific knowledge. Specifically, this chapter focuses on changes in life-forms likely to result from management actions. The relationship between research and management and the influence of land tenure on resource management are briefly outlined to provide the context for discussing manipulation of woody plants and grasses.

Linking Research and Management

Ecologists have generally failed to conduct experiments relevant to managers (Underwood 1995), and managerial agencies often resist criticisms of performance or suggestions for improvement (Longood and Simmel 1972, Underwood 1995). These factors have contributed to poorly developed and sometimes adversarial relationships between managers and scientists. To address this problem, scientists should be proactive, rather than reactive, with respect to resource management issues, and managers should be familiar with scientific principles. Chapter 7 briefly discusses the scientific principles particularly relevant to management and describes the responsibilities of researchers with respect to management.

The realm of science represents a substantial reservoir of relatively untapped information available to resource managers (McPherson and Weltzin 1997). Managers in need of scientific information are encouraged to use existing data, work closely with the scientific community, and communicate the need for specific information. Scientists seek generality in their research, which is contradictory to the site-specific information needed by managers. Thus, it becomes incumbent on managers to determine which elements of the expansive ecological literature are relevant to specific management objectives. In addition, managers must be able to extrapolate concepts generated elsewhere into activities at a particular site.

Finally, resource managers must understand how scientific knowledge is obtained. For example, not all information generated by scientists will enable managers to accurately predict the response of a plant community to a specific disturbance or manipulation. Some research findings present untested hypotheses rather than documented responses of different ecosystems to different disturbances. In fact, most research journals encourage authors to present potential mechanisms for observed patterns. Managers and policy makers routinely confuse these tentative, untested hypotheses for tested, documented phenomena and use the former as a basis for decisions. When asked to make predictions, the usual response of scientists is to blithely proceed; if they perceive the problem inherent in making predictions based on tentative hypotheses, they equivocate. Neither blithely proceeding nor equivocating provides useful information for solving management problems.

Land Tenure and Policy Issues

Patterns of land ownership have marked effects on policy and land management at regional scales. Public lands are subject to complex laws, regulations, and guidelines, often with an overall goal of managing for multiple uses. Conservation of soils and native vegetation on public lands often is a higher priority than economic viability. In fact, the idea of returning to "pristine" conditions as the ultimate goal of resource management is pervasive (Brown and MacLeod 1996) despite the inability of managers or scientists to describe these conditions and the inability of these conditions to satisfy societal demands. In contrast, most private lands must be managed in an economically efficient manner. Thus, common goals on private lands include maximizing forage production for livestock or wildlife, or developing resource-based recreational opportunities.

Preservation of ponderosa pine, pinyon-juniper, mesquite, and southwestern oak savannas in the United States is largely ensured by federal ownership of most of these savannas. Forest Service activities are continually judged in reference to conservation of the native vegetation (Allen 1989), and that practice probably will continue. However, residential development of privately owned areas adjacent to public savannas and the possibility of land sales to private owners by states in the western United States will threaten the continuity of savanna landscapes by further constricting large-scale processes such as fire, patterns of water drainage, and movement of animals. The ramifications of these activities are unclear but may include decreases in the diversity and production of plants and animals.

Preservation of Californian oak, longleaf pine, and the few remnant midwestern oak savannas is largely dependent on private enterprise because only small portions of these communities are in the public domain. Preservation of these savannas is increasingly facilitated by partnerships between private landowners and public agencies, as well as grassroots efforts that focus on education and mitigation of development-induced impacts (e.g., Antunez de Mayolo 1991, Giusti et al. 1991, Botts et al. 1994, Landers et al. 1995). In addition, conservation easements are being used to negate increased taxes and reduce urbanization (e.g., McDonald 1995).

Effects of land tenure on preservation of Mexican oak savannas are unknown. Most land is managed communally (see chapter 1), and population growth has been rapid (see chapter 5): these factors have led to high, but unquantified, rates of resource extraction. The virtual absence of data on historical and current land uses in Mexico precludes clear identification—much less the development of recommendations—of relevant land tenure and policy issues.

Overstory-Understory Relationships

The relationship between woody plants and grasses is generally inverse (see chapter 2, esp. fig. 2.3). The relationship may be linear on sites that have been protected from livestock grazing but is logarithmic on most sites (McPherson 1992a). The logarithmic relationship between overstory and understory plants indicates that at low levels of woody plant abundance, a small increase in woody plant abundance produces a relatively large decrease in grass abundance, while at higher levels of cover or density, additional increases in woody plant abundance produce little change in grass abundance. The logarithmic relationship has been attributed to the accrual of a competitive advantage to woody plants as a result of livestock grazing and fire exclusion (Walker et al. 1986, McPherson 1992a): if woody plant seedlings survive beyond the seedling stage, they are able to usurp resources (e.g., light, soil moisture) to a greater extent than grasses.

The inverse relationship between woody plants and grasses complicates the search for an "optimal" solution for savanna management. Objectives that require high levels of grass biomass (e.g., some types of livestock production) conflict with objectives that require high levels of woody plant biomass (e.g., fuelwood production). Conflicting objectives must be addressed at relatively fine spatial scales because overstory effects are highly localized and may be restricted to the area immediately beneath the woody plant, the precise relationship between woody plants and grasses varies in a site-specific manner, and management is by necessity site- and objective-specific.

Factors Affecting the Management of Savannas

The remainder of this chapter focuses on factors that are likely to affect the relative proportion of woody plants and grasses. These include several managerial tools, the use of which includes the following assumptions: tradeoffs have been considered and acknowledged, and the optimal mix of life-forms has been identified, consistent with management goals and objectives. Restoration and maintenance of savanna physiognomy will require decreased abundance of woody plants in some savannas (e.g., most longleaf pine, ponderosa pine, pinyon-juniper, southwestern oak, and mesquite savannas) and increased abundance of woody plants in others (e.g., many midwestern and Californian oak savannas). These goals can be achieved by manipulating populations of woody plants, grasses, or both. Propagation of dominant woody plants is relatively simple and straightforward (Young and Young 1992); oaks represent an exception because acorns are universally difficult to store and handle (e.g., Bonner and Vozzo 1987, Young and Young 1992) and seedlings are susceptible to high mortality immediately after their emergence in the field (see chapter 3).

Livestock Grazing

Livestock grazing is a potentially powerful tool for altering vegetation. Cattle generally enhance establishment and growth of woody plants at the expense of grasses (see chapter 3). However, grazing by goats and horses may favor grasses because these animals select woody plants to a greater extent than do other livestock (Vallentine 1990, Stuth 1991).

Many land management activities are designed to enhance grass production so that the additional forage can be converted to animal products (e.g., meat, wool, leather). Livestock grazing is one of the few management activities in which the product (livestock) is also the tool for vegetation management. Unfortunately, the effects of livestock grazing are generally opposite those desired by livestock producers: most livestock grazing contributes to increased woody plant abundance at the expense of grasses.

Trial and error has been the dominant method used to determine stocking rates and grazing systems. This is not surprising, considering the site-specificity of edaphic factors, species composition, and cultural features (McPherson and Weltzin 1997). Soil surveys and range site guides produced by agencies of the United States government are detailed sources of information regarding site-specific management recommendations. Publications of the USDA Natural Resource Conservation Service (formerly the Soil Conservation Service) are particularly noteworthy. The USDA Forest Service and USDI Bureau of Land Management have also produced management guidelines that are relatively site-specific and are therefore widely recommended for resource managers. Although these publications provide considerable information that may be useful for management, they are based on a fundamentally flawed model of vegetation dynamics (Clements 1916, Dyksterhuis 1949, Parker 1954). This model views vegetation change as a simple, linear process that culminates in one stable state. Thus, this model does not accommodate the kinds of changes in land use, disturbance regimes, and climate that typify North American savannas (Westoby et al. 1989, Svejcar and Brown 1991). A more appropriate model of vegetation dynamics, the state-and-transition model, has been proposed (Westoby et al. 1989) but not implemented (e.g., Joyce 1993, Scarnecchia 1995).

Consistent with current practices, future livestock activities in North American savannas will be dictated by site-specific edaphic, ecologic, and cultural factors. In addition, social, economic, and political factors will become increasingly important determinants of the role and importance of livestock grazing, particularly on public lands. Livestock managers who anticipate and respond to ecosystems altered by changing climates, fire regimes, and social and political attitudes should fare well compared to inflexible managers (McPherson and Weltzin 1997).

The current role of livestock grazing as a regulator of vegetation change may be considerably reduced compared to its historic role (McPherson and Weltzin 1997), with the possible exception of grazing in Californian oak savannas. Livestock exclusion often will not prevent, and may not delay, a substantial increase in the abundance of many woody plants, as evidenced by numerous exclosure studies in the western United States (Brown 1950, Gardner 1951, Glendening 1952, Smith and Schmutz 1975, Smeins et al. 1976, Wright 1982, Hennessey et al. 1983, Roundy and Jordan 1988, Brady et al. 1989, McClaran et al. 1992). Results of these and other studies suggest that many former grasslands and savannas have crossed an ecological threshold to stable domination by closed-canopy woodlands (West 1984, Jameson 1987, Archer 1989, McPherson and Wright 1990a, Covington and Moore 1994a, 1994b, Covington et al. 1994). Furthermore, stocking rates and grazing systems currently in use are designed to minimize impacts to vegetation. Thus, although livestock grazing (particularly in combination with other factors) contributed to broad-scale vegetation changes shortly after Anglo settlement, excluding livestock from most sites now will have little impact on the abundance of woody plants during the next several decades. Finally, exclusion of livestock from most public savannas has proved politically infeasible (Brown and McDonald 1995). Some Californian oak savannas are exceptional in terms of livestock impacts: phenological differences between oaks and nonnative annual grasses (see chapter 3) make oaks particularly susceptible to livestock grazing, so recruitment of oak trees may be enhanced by excluding livestock.

Management of Native Herbivores

Native herbivores can be managed to affect the relationship between woody plants and grasses. Although the effects of several herbivores on community structure and function are largely unknown (e.g., deer, pronghorn, javelina, squirrels), the effects of

Figure 6.1 Prairie dogs were extirpated from many North American savannas because they were thought to compete with cattle for forage. However, these native herbivores may constrain the abundance of woody plants. (Photograph courtesy of Heather Germaine.)

other native herbivores (e.g., prairie dogs) have been well documented (e.g., Huntly and Inouye 1988, Whicker and Detling 1988, Brown and Heske 1990, Huntly 1991).

Populations of black-tailed prairie dogs *(Cynomys ludovicianus)* were virtually eliminated from the United States because these rodents were thought to compete with cattle for forage (Miller et al. 1990, 1994). Prairie dogs are still relatively common in Mexico (Ceballos et al. 1993, Felger and Wilson 1995). They have profound effects on microtopography, soils, plants, and other animals (Huntly and Inouye 1988, Whicker and Detling 1988, Miller et al. 1994), and regional extirpation of the species may have contributed to rapid increases in woody plant abundance (see chapter 3).

Similarly, native herbivores in other savannas may constrain woody plant establishment (e.g., Reynolds and Glendening 1949, Griffin 1971, Evans 1987, Borchert et al. 1989). In particular, pocket gophers (Family Geomyidae) may affect woody plant survival and ecosystem properties in a manner similar to that of prairie dogs.

Reintroduction of native herbivores, particularly prairie dogs, may represent a cost-effective strategy for reducing woody plant abundance (fig. 6.1). It should be noted that the role of this once-common herbivore in reducing the abundance of woody plants such as ponderosa pine, longleaf pine, and oaks has not been determined. Furthermore, prairie dogs may have undesirable effects on aboveground net primary production or species composition of herbs. However, well-controlled reintroduction of these native herbivores should elucidate their potential role in the control of woody plants (McPherson and Weltzin 1997). Prairie dogs can be easily extirpated, which will allow rapid termination of experimental reintroduction.

If increasing woody plants is a management goal, herbivory by native and nonnative animals may need to be reduced. For example, restoration of Californian oak savannas is often dependent upon exclusion of small mammals and invertebrates from individually protected oak seedlings (Adams et al. 1992). Several styles of plastic tubes have been developed and marketed specifically for protecting oak seedlings. Similarly, local exclusion of feral hogs, which can destroy thousands of seedlings each day, may be necessary to ensure survival of longleaf pine seedlings (Hine 1925, Wahlenberg 1946). These practices, though expensive, are usually effective for increasing recruitment of woody plants.

Fire Management

Naturally occurring fires were a frequent phenomenon in North American savannas before Anglo settlement (see chapter 4). After several decades of fire suppression, there is increasing interest in restoring fire to these systems (e.g., Wright 1974, Streng and Harcombe 1982, Wright and Bailey 1982, Steuter and Wright 1983, Swetnam 1990, Covington and Moore 1992, 1994a, McClain et al. 1993, Allen 1995, Landers et al. 1995). However, simple recommendations concerning fire management cannot be developed, because objectives vary between sites and interactions between fire and the environment result in strongly site-specific responses of vegetation to fire (see chapter 3). In addition, fragmented patterns of land ownership preclude restoration of fire regimes on landscape scales. Finally, although periodic fires may be necessary to maintain savanna physiognomy, the associated costs (e.g., instantaneous loss of herbaceous forage, short-term increase of atmospheric smoke, and altered wildlife habitat) may preclude the reintroduction of fire in many areas. The following sections briefly describe the implications of continued fire suppression and of reintroducing prescribed fires.

Fire Suppression. Fire suppression has been implicated in the increased density and cover of woody plants in most North American savannas since Anglo settlement (see chapter 4). Continued fire suppression in most savannas will probably increase fuel loads and therefore increase the probability of wildfires (Steuter and McPherson 1995). Furthermore, contemporary fires that consume these high fuel loads tend to be high-intensity, stand-replacing events that contrast sharply with the low-intensity surface fires of presettlement savannas (Covington and Moore 1994a, MacCleery 1995, Swetnam and Baisan 1995). In contrast to the increased probability of fires associated with fire suppression in most savannas, fire suppression in southeastern pine and mesquite savannas may reduce the probability of subsequent fires because of fine fuel loads that have decreased under overstory canopies (see chapter 3). In either case, continued fire suppression decreases management flexibility over time: the feasibility of using prescribed fire as a management tool declines as the abundance of woody plants increases. Therefore, fire suppression is not generally recommended except as a means of preserving valuable cultural resources.

Prescribed Fires. Fires are generally perceived as more "natural" and hence environmentally more appropriate than other vegetation-management tools, presumably because fires were a frequent component of presettlement landscapes. In addition, prescribed fires are less expensive to implement than are other manipulations (Wright and Bailey 1982, Vallentine 1989). However, prescribed fires are difficult to apply and control. Also, state-issued permits must be secured, and prescribed fires generate atmospheric pollutants and cause alterations in habitat that may be undesirable for some species of wildlife. Furthermore, prescribed fires may interfere with other management practices. For example, management

Figure 6.2 Prescribed fires are used to reduce woody plant cover in North American savannas.

of livestock must accommodate patterns of prescribed fire application: livestock must be removed from savannas long enough to allow development of a continuous stand of herbaceous fuels. This outcome may require over two years on some sites and may never occur on others. Postfire recovery of herbaceous biomass may require an additional year (Wright and Bailey 1982, Vallentine 1989). Also, prescribed fires must be applied during a time of year that will minimally affect critical life-history stages if production of native animals is a goal. Thus, longleaf pine savannas usually are prescribe-burned in winter to accommodate the nesting of bobwhite quail (Stoddard 1931, Rosene 1969, Landers and Mueller 1985).

If control of woody plants is a primary management objective, then implementation of periodic prescribed fires should be considered (fig. 6.2). Many specific issues should be evaluated before embarking on a program of prescribed burning (Wright and Bailey 1982, Vallentine 1989). For example, fires designed to decrease mesquite density may instead stimulate resprouting, thereby increasing stem density. Thus, knowledge of species-level response is fundamental to the implementation of a successful program.

In general, frequent fires are detrimental to woody plants and beneficial to short-lived herbs, especially annuals. However, periodic fires may be insufficient to control the abundance of woody plants on many sites. Rather, fire management must be integrated with appropriate grazing management. In many cases, major cultural inputs (e.g., mechanical or chemical treatments) may be required to create and maintain savannas because lack of fine fuels precludes the use of fire as a restorative technique. This is particularly problematic in former savannas dominated by closed-canopy stands of woody plants (Wright and Bailey 1982). In addition, once savanna physiognomy is established, maintenance of the savanna will require a site-specific fire regime. For example, savanna maintenance requires a higher fire frequency on deep-soiled, highly productive sites than on erosion-prone or shallow-soiled sites (Steuter and McPherson 1995).

Mechanical and Chemical Treatments

Many methods of mechanically and chemically controlling plant populations have been developed in association with the range and forestry professions. The efficacy of these treatments is strongly dependent on environmental conditions and on plant morphology and phenology during and after application. Furthermore, application of these treatments requires considerable skill and, in the case of some herbicides, licenses or permits. (For a more detailed discussion of these methods, readers are referred to relevant publications such as Smith 1986, Vallentine 1989, Luken 1990 and agencies such as state pesticide information and training offices and county extension services).

Most mechanical and chemical treatments are designed to reduce the abundance of specific woody plants. The large number and variety of techniques allow considerable flexibility with respect to size, growth-form, and density of target species. Nonetheless, there is considerable variability in the efficiency of treatments. Furthermore, high economic costs associated with mechanical and chemical methods impose significant constraints:

they are generally either labor-intensive or dependent on large investments in capital and petroleum-based fuels.

Mechanical and chemical treatments can be applied to individual plants (i.e., "spot" control) or to entire stands (i.e., "broadcast"). Spot control is commonly recommended when target plants are widely scattered; in contrast, dense stands of plants are treated more efficiently using broadcast methods (Vallentine 1989).

Individual plant treatments are labor-intensive because each target plant must be visited and treated. The efficacy of this method is usually excellent, for the same reason. Mechanical control of individual plants includes cutting or chopping, girdling, and whole-plant excavation. The latter treatment is usually applied with one of several specialized attachments on a tractor. Herbicides may be applied to individual plants by injection into the stem or by spraying onto leaves or stems. Adequate control of some species requires a combination of mechanical and chemical treatments (e.g., herbicides applied to stems immediately after they are cut or girdled; Vallentine 1989).

Whole-stand treatments require the use of expensive capital equipment (e.g., tractors, bulldozers, aircraft). Mechanical methods are designed to sever or uproot plants and are usually accomplished with specially adapted bulldozers. A common technique in high-density stands of woody plants involves dragging an anchor chain or cable between two bulldozers. The chain or cable, which is often modified with steel rails or disks, severs or uproots woody plants. Chaining and cabling are especially effective for controlling nonsprouting species with brittle stems. Adequate control of species with more flexible stems, especially those that resprout, generally requires plowing, disking, or use of a specialized root plow.

Because of their expense and the perception that they are environmentally inappropriate, the use of mechanical and herbicidal methods has declined in the last few decades. However, when used properly, these methods represent an important and appropriate tool for reducing woody plant abundance in closed-canopy stands. Thus, restoration and maintenance of savanna physiognomy often is accomplished most appropriately by using a combination of techniques. Grazing management, fire management, and use of chemical and mechanical treatments should be considered carefully when managing savannas.

Climate Change

Managers must recognize the potential effects of ongoing and impending changes in atmospheric and climatic factors on current and future vegetation (see chapter 5). This is perhaps the most critical factor, second only to urbanization, facing land managers who are concerned about the long-term structure and function of North American savannas. Individual managers have little or no control over changes in climate, which parallels their influence on urbanization. Nonetheless, proactive managers will want to know the probability of various climate scenarios and subsequent vegetation responses within a specific region. This information can be determined with a combination of models and experiments.

Changes in atmosphere and climate are important to resource managers because (1) directional changes in climate are virtually certain to continue, a trend that will have profound impacts on vegetation; (2) climatically induced effects will occur over larger areas and have potentially greater impacts than changes in management practices; and (3) the influence of management practices on vegetation will be difficult to predict or interpret without explicit consideration of relatively rapid changes in atmospheric and climatic factors. Accurate prediction of ecosystem response to atmospheric and climatic changes will facilitate proactive resource management. Conversely, if managers are unaware of potential and ongoing changes in atmospheric and climatic conditions (i.e., means and extremes) and their associated effects on ecosystems, efforts to effectively manage those ecosystems could be severely handicapped.

Rising atmospheric concentrations of CO_2 and other trace gases have the potential to replace traditional disturbances (i.e., livestock grazing and fire suppression) as important regulators of vegetation change (McPherson and Weltzin 1997). It is unlikely that presettlement fire regimes or livestock grazing regimes will be restored in the foreseeable future, and even if they were, woody plant abundance would not change on many sites without major cultural inputs. Although past increases in $[CO_2]_{atm}$ may have had no appreciable impact on savanna vegetation (Archer et al. 1995), continued increases in $[CO_2]_{atm}$ may contribute to an increased abundance of woody plants in grasslands and savannas, particularly if these increases are accompanied by shifts in the seasonality of precipitation or changes in other climatic factors. The implications for management are clear: if increased atmospheric CO_2 concentrations or associated changes in precipitation patterns enhance recruitment of woody plants, then resource managers must consider climatic changes when developing long-range management goals and objectives. Alternatively, if changes in $[CO_2]_{atm}$ and precipitation do not affect woody plant recruitment, or if changes that occur are desirable, then management actions may be superfluous.

Long-term meteorological records indicate that periodic droughts occur frequently in all North American savannas. In fact, for most meteorological stations, annual precipitation is below the long-term average during at least 60% of the years since monitoring began. Management actions should account for episodic and unpredictable precipitation patterns. For example, stocking rates of livestock should be based on levels of herbaceous production associated with years of below-average precipitation. Resource managers should respond to increased herbaceous production associated with above-average precipitation by shifting grazing patterns (e.g., by using stocker animals or altering patterns across elevational gradients) or by using the additional forage as fuel in prescribed fires. The importance of management flexibility is very high and will continue to increase in concert with the expected increase in climatically extreme events (see chapter 5).

Summary

Appropriate management can be prescribed only after goals and objectives are clearly defined. After this initial step, knowledge of species life-histories, plant-animal interactions, and response of ecosystems to land uses, disturbance, and environment may facilitate the prediction of a given system's response to management actions. Some of this information has been generated by previous research, as described in chapters 3–6. Additional research, designed to facilitate management through development of general principles and determination of mechanisms underlying observed changes, is described in the following chapter. However, the role of knowledge derived from personal experience should not be underestimated: because management is necessarily site- and objective-specific, decisions should be made by those managers who are knowledgeable about, and responsible for, individual sites.

7

Research Needs

> *Children and scientists share an outlook on life.* If I do this, what will happen? *The unfamiliar and the strange—these are the domain of children and scientists.*
> *James Gleick,* Genius: The Life and Science of Richard Feynman

A considerable body of research has been developed through investigation of the structure and function of North American savannas. This research, described in the preceding chapters, has been instrumental in the determination of biogeographical, biogeochemical, environmental, and physiological patterns that characterize savanna ecosystems. In addition, research has elucidated some of the underlying mechanisms that control patterns of species distribution and abundance. Most important, however, research to date has identified many tentative explanations (i.e., hypotheses) for observed ecological phenomena. Few of these hypotheses have been explicitly tested, which has limited the ability of ecology, as a discipline, to foresee or help solve managerial problems (Underwood 1995). The contribution of science to management is further limited by the lack of conceptual unity within ecology and the disparity in the goals of science and management, as described below.

The unique characteristics of each type of savanna, even within North America, impose significant constraints on the development of parsimonious concepts, principles, and theories. This lack of conceptual unity is widely recognized in ecology (Keddy 1989, Peters 1991, Pickett et al. 1994) and natural resource management (Underwood 1995). The paucity of unifying principles imposes an important dichotomy on science and management: general concepts, which science should strive to attain, have little utility for site-specific management. Yet detailed understanding of a particular system, which is required for appropriate management, makes little contribution to ecological theory. This disparity in goals is a significant obstacle to relevant discourse between science and management (see chapter 6).

Management and conservation are ultimately conducted at the local level. As a result, management decisions must be temporally and spatially finite and should be goal- and objective-specific; they should be made by those managers most familiar with local systems. Ecologists can make the greatest contribution to management and conservation by addressing questions that are relevant to resource management and by focusing their research activities at appropriate temporal and spatial scales (Allen et al. 1984). I suggest that these scales are temporally intermediate (i.e., ranging from years to decades) and spatially local (e.g., square kilometers). Of course, contemporary ecological research should be conducted within the context of the longer temporal scales and greater spatial scales at which policy decisions are made. For example, experimental research on interactions between climate and vegetation should be conducted within individual savannas for periods of a few years, but the research should be couched within patterns and processes observed at regional to global spatial scales and decadal to centennial temporal scales. In other words, the context for ecological experiments should be provided by a variety of sources, including observations, management issues (McPherson and Weltzin 1997), long-term databases (Likens 1989, Risser 1991), cross-system comparisons (Cole et al. 1991), and large-scale manipulations (Likens 1985, Carpenter et al. 1995, Carpenter 1996).

Two examples amenable to this type of approach are midwestern oak savannas and pinyon-juniper savannas, represented in the Long-Term Ecological Research (LTER) program administered by the National Science Foundation at Cedar Creek Natural History Area and Sevilleta National Wildlife Refuge, respectively. These sites, which are among the newest members of the LTER network, should provide a significant source of long-term data on these savannas for the foreseeable future.

The Role of Science in Resource Management

As with any human endeavor, science shares many characteristics with everyday activities. For example, observations of recurring events are used to infer general patterns in banking and fishing, as well as most scientific disciplines. The discussion herein focuses on features that are unique to science. I assume in this chapter that science is obliged in part to offer explanatory and predictive power about the natural world. An additional assumption is that the scientific method, which includes explicit hypothesis-testing, is among the most efficient and valid techniques for acquiring reliable knowledge. The scientific method should be used to elucidate mechanisms underlying observed patterns; such elucidation is the key to predicting and understanding natural systems (Levin 1992; but see Pickett et al. 1994).

Resource managers need reliable scientific information to effectively manage plant communities and ecological processes. Because abundant data are available in a wide variety of qualities (cf. Huxley 1932), managers must extract relevant information from the body of knowledge to address management decisions. Additional factors contribute to the dilemma that managers face as they attempt to incorporate scientific knowledge into management decisions: much of the available information is contradictory or inconsistent, and many scientists still attempt to provide mechanistic explanations about ecosystem function based on descriptive research. This latter tendency has trapped scientists into making predictions about things they cannot predict (Peters 1991, Underwood 1995). Adherence to scientific principles, including hypothesis-testing, will improve communications between resource managers and scientists while increasing the credibility of both groups.

From a modern scientific perspective, a hypothesis is a candidate explanation for a pattern observed in nature (Medawar 1984, Matter and Mannan 1989). That is, a hypothesis is a potential reason for the pattern; it should be testable and falsifiable (Popper 1981). Hypothesis-testing is a fundamental attribute of science that is absent from virtually all other activities. Science is a process by which competing hypotheses are examined, tested, and rejected. Failure to falsify a hypothesis with an appropriately designed test is interpreted as confirmatory evidence that the hypothesis is accurate.

A hypothesis is not merely a statement likely to be factual, which is then "tested" by observation. If we accept any statement (e.g., those involving a pattern) as a hypothesis, then the scientific method need not be invoked—we can merely look for the pattern. Indeed, if observation is sufficient for the development of reliable knowledge, then science has little to offer that goes beyond the information provided by everyday activities. Much ecological research is terminated after discovery of a pattern; the cause of the pattern is not determined (Romesburg 1981, Willson 1981).

Use of sophisticated technological (e.g., microscopes) or methodological (e.g., statistics) tools does not imply that hypothesis-testing is involved, if these tools are used merely to detect a pattern. Pattern recognition (i.e., assessment of statements likely to be factual) often involves significant technological innovation. In contrast, hypothesis-testing is a scientific activity that need not involve state-of-the-art technology. Some evidence suggests that even scientific organizations fail to distinguish between technological and scientific advances. For example, scientific accolades are usually reserved primarily for individuals who pioneer technological advances instead of being given to those who make significant scientific contributions (Dulbecco 1995).

Fragmentation within the ranks of ecology seriously threatens the ability of ecology as a discipline to solve important environmental problems. The divide between scientists and managers remains vast, despite increasingly frequent pleas for ecologists to focus on managerial problems (e.g., Keddy 1989, Kessler et al. 1992, Sharitz et al. 1992, Underwood 1995). In addition, a rift between theoretical ecologists and applied ecologists developed rapidly after circa 1960. The divergence of viewpoints corresponded to a dramatic increase in the number of applied ecologists (e.g., foresters, range scientists, wildlife ecologists) as natural resource programs expanded at universities and government agencies. Most applied ecologists addressed problems that were viewed by theoretical ecologists as site-specific or poorly grounded in ecological theory. In contrast, theoretical ecologists have increasingly distanced themselves from applied disciplines by ignoring the historical and contemporary role of humans in ecological processes, by downplaying the importance of social, political, and economic processes on ecological processes, and by discounting or ignoring literature associated with applied disciplines (e.g., forestry, wildlife management, range science). Thus, theoretical ecologists have "discovered" various phenomena years or decades after these phenomena were accepted by the management community and have tended to conduct research that is viewed as irrelevant by natural resource managers.

Testing Ecological Hypotheses

Some ecologists (exemplified by Peters 1991) have suggested that ecology makes the greatest contribution to solving management problems by developing predictive relationships based on correlations. This view suggests that ecologists should describe as many patterns as possible, without seeking to determine underlying mechanisms. An even more extreme view is described by Weiner (1995), who observed that considerable ecological research is conducted with no regard to determining patterns or testing hypotheses. In contrast to these phenomenological viewpoints, most ecologists subscribe to a central tenet of the modern philosophy of science: determining the mechanisms underlying observed patterns is fundamental to understanding and predicting ecosystem response to changes in the physical or biological environment, and therefore is necessary for improving management (e.g., Hairston 1989, Matter and Mannan 1989, Levin 1992, Weiner 1995; but see Pickett et al. 1994).

Because hypotheses are merely candidate explanations for observed patterns, they should be tested. Experimentation (i.e., artificial application of treatment conditions followed by monitoring) is an efficient and appropriate means for testing hypotheses about ecological phenomena; it is often the only means for doing so (Simberloff 1983, Campbell et al. 1991). Experimentation is necessary for disentangling important driving variables, which may be correlated strongly with other factors under investigation

(Gurevitch and Collins 1994). Identification of underlying mechanisms of vegetation change enables prediction of vegetation response to changes in driving variables with levels of certainty and on spatial and temporal scales useful to resource managers.

In contrast to the majority of ecologists, most managers of ecosystems do not understand that experiments are necessary to determine mechanisms of vegetation change. In the absence of experimental research, managers and policy makers must rely on the results of descriptive studies. Unfortunately, these studies often produce conflicting interpretations of underlying mechanism and are plagued by weak inference (Platt 1964): descriptive studies (including "natural" experiments, *sensu* Diamond 1986) are forced to infer mechanism based on pattern. They are therefore poorly suited for determining underlying mechanisms or causes of patterns because there is no test involved (Popper 1981, Keddy 1989). Even rigorous, long-term vegetation monitoring is incapable of revealing causes of vegetation change because of confounding between the many factors that potentially contribute to shifts in species composition (e.g., Wondzell and Ludwig 1995).

Examples of "natural" experiments abound in the ecological literature, but results of these studies should be interpreted judiciously. For example, savanna researchers have routinely compared recently burned (or grazed) areas to adjacent unburned (ungrazed) areas and have concluded that observed differences in species composition were the direct result of the disturbance under study. Before reaching this conclusion, it is appropriate to ask why one area burned while the other did not. Preburn differences in productivity, fuel continuity, fuel moisture content, plant phenology, topography, or edaphic factors may have caused the observed fire pattern. Because these factors influence and are influenced by species composition, they cannot be ruled out as candidate explanations for postfire differences in species composition.

Although descriptive studies are necessary and important for describing savanna structure and identifying hypotheses, reliance on this research approach severely constrains the ability of ecology to solve managerial problems. In addition, the poor predictive power of ecology (Peters 1991) indicates that our knowledge of ecosystem function is severely limited (Stanley 1995). Inability to understand ecosystem function and unjustified reliance on descriptive research are among the most important obstacles that prevent ecology from making significant progress toward solving environmental problems and from being a predictive science. Many ecologists (e.g., Hairston 1989, Keddy 1989, Collins and Gurevitch 1994) have concluded that field-based manipulative experiments represent a logical approach for future research.

Results of ecological research are likely to be highly site-specific (Keddy 1989, Tilman 1990), and it is infeasible to conduct experiments in each type of soil and vegetation. Therefore, experiments should be designed to have the maximum possible generality to other systems (Keddy 1989). For example, the pattern under investigation should be widespread (e.g., shifts in physiognomy), selected species should be "representative" of other species (e.g., of similar life-form), the factors manipulated in experiments should have broad generality (e.g., biomass), experiments should be arranged along naturally occurring gradients (e.g., soil moisture, elevation), and experiments should be conducted at spatial (e.g., plant community) and temporal (i.e., annual or decadal) scales appropriate to management of plant communities.

Experiments need not be conducted at small spatial scales. For example, ecosystem-level experiments (i.e., relatively large-scale manipulation of ecosystems) represent an important, often-overlooked technique that can be used to increase predictive power and credibility in ecology. Ecosystem-level experiments may be used to bridge gaps between small-scale experiments and uncontrolled observations, including "natural" experiments. However, they are difficult to implement and interpret (Carpenter et al. 1995, Lawton 1995): they require knowledge of natural histories of species, an understanding of natural disturbances, and considerable foresight and planning. Fortunately, ecology has generated considerable information about the natural history of dominant species and natural disturbances in many ecosystems. Similarly, foresight and planning should not be limiting factors in scientific research. Time and money will continue to be in short supply, but this situation will grow more serious if ecology does not establish itself as a source of reliable knowledge about environmental management (Peters 1991, Underwood 1995).

In addition to posing questions that are relevant to resource management and that investigate mechanisms, scientists should be concerned with the development of research questions that are tractable. Asking why woody plants or savannas are present at a particular place and time forces the investigator to rely on correlation. By contrast, asking why woody plants or savannas are not present (e.g., in locations that appear suitable) forces the investigator to search for constraints, and therefore mechanisms. Although Harper (1977, 1982) presented a compelling case for tractable, mechanistic research focused on applied ecological issues nearly two decades ago, the underwhelming response by ecologists indicates that his message bears repeating.

The remainder of this chapter outlines research needs in North American savannas, with a strong focus on changes in life-form. In particular, understanding the factors that influence the relative proportion of woody plants and grasses is fundamental to predicting response of savannas to management, disturbance, and environmental change. The research outlined herein is intended to enable managers to better predict responses of savannas to these factors.

Experiments for Savannas

Many gaps in our knowledge of savanna ecology are identified in previous chapters. For example, chapter 3 illustrates that there is insufficient knowledge of (1) effects of native herbivores on the relative proportion of woody plants and grasses; (2) indirect effects of livestock grazing on habitat for native herbivores, and subsequent effects on interactions between woody plants and grasses; (3) effects of gap size on establishment of woody plants; (4) the role of soil moisture and temperature on recruitment and mortality of woody plants and grasses; (5) effects of short-term or episodic climatic events on vegetation; (6) soil-vegetation interactions; and (7) interactive effects of herbivory, fire, climate, and soils. To avoid redundancy, this section does not provide a comprehensive list of research needs. Rather, it contains examples of the kinds of research that may facilitate appropriate management of North American savannas.

Management Manipulations

Extensive research on grazing management (reviewed by Vallentine 1990, Heitschmidt and Stuth 1991) and ex post facto woody plant control (reviewed by Vallentine 1989) has produced a considerable literature on management manipulations that serves as a basis for site-specific recommendations for rates and timing of livestock grazing, prescribed fire frequency, and woody plant control. However, it is clear that research on these factors cannot replace practical land-management experience. Therefore, continued research on traditional management manipulations is recommended only within the context of likely changes in atmospheric and climatic conditions.

Effects of native herbivores are poorly known and little studied but of great potential importance. Because native herbivores may impose significant constraints on establishment of woody plants, research should focus on potential effects of different animal species on establishment and persistence of woody plants. Limited research in North American savannas suggests that vertebrates and invertebrates are seasonally important constraints on survival of woody plant seedlings (e.g., Griffin 1971, Adams et al. 1992, McPherson 1993, Germaine 1997). However, the spatial extent and importance of these mortality vectors have not been determined, let alone the particular species responsible for the observed mortality. Invertebrate herbivory may increase dramatically in a CO_2-enriched world as a result of reduced forage quality (see chapter 5).

The effects of different frequencies and seasons of fires on the structure and function of savannas are largely unknown. Research on fire effects should focus on the influence of different fire seasons and frequencies on physiognomy, species composition, and nutrient cycling. Recent research in longleaf pine savannas (Platt et al. 1988a, 1991, Streng et al. 1993) sets a high standard for studying fire ecology in other savannas. Additional components that should be considered include fire intensity, variability in fire regimes, and interactions between fire and climate. One product of this research will be increased understanding of the relationship between ecosystem structure (e.g., relative proportion of woody plants and grasses) and ecosystem function (e.g., nitrogen cycling, productivity, carbon flux).

Atmospheric and Climatic Change

The greatest gap in our knowledge of North American savannas is the potential effect of atmospheric and climatic change on vegetation interactions and community- and ecosystem-level processes. Because savannas are ecotonal between grasslands and forests, they may be particularly sensitive to changes in atmospheric and climatic conditions; thus, additional research in these systems is needed (Mooney et al. 1991). Currently, prediction of future changes in distribution and composition of plant communities is difficult, given the background of recurrent disturbances and the complexity and paucity of knowledge about regionally specific climate change (Mitchell et al. 1990). However, the determination of most-likely scenarios of atmospheric and climate change is relatively straightforward, and these scenarios may be tested experimentally at scales appropriate to resource management (Kingsolver et al. 1993, Mooney and Chapin 1994, Koch and Mooney 1996). Experiments that focus on interactions between various atmospheric factors (e.g., concentrations of greenhouse gases), climatic factors (e.g., temperature, precipitation), and disturbance regimes likely will provide the greatest simultaneous contribution to ecological theory and resource management (McPherson and Weltzin 1997).

Although experiments on atmospheric and climatic change should be focused on management applications, they should be well grounded in sound ecological theory. For example, soil resource partitioning is widely invoked to explain the apparent long-term stability of savannas (e.g., Walter 1954, 1979, Knoop and Walker 1985, Sala et al. 1989, Brown and Archer 1990, Bush and Van Auken 1991, Skarpe 1992). Bimodal patterns of precipitation distribution are thought to allow the stable coexistence of woody plants and grasses: shallow-rooted grasses use precipitation that falls during the growing season, whereas deep-rooted woody plants use moisture that percolates through surface soil layers when grasses are dormant (Neilson 1986, Archer 1989). Although this hypothesis is intuitively logical, it has yet to be explicitly tested. Implications for the management of plant communities are profound. If resource partitioning does occur, then shifts in precipitation seasonality that are predicted to occur as a result of atmospheric CO_2 enrichment and changes in water-use efficiency between woody C_3 and herbaceous C_4 plants may affect soil moisture pools. In turn, changes in soil moisture may affect interactions between woody plants and grasses where they currently coexist or may allow one or the other life-form to establish where it is currently excluded by environmental constraints. If the seasonal distribution of precipitation changes within the next few decades, rapid and dramatic changes in the relative proportion of woody plants and grasses may occur. These structural changes would have important implications for ecosystem function. Experimentation will enable prediction of not only the relative importance of current precipitation regimes on proportions and distributions of woody plants and grasses, but also potential effects of changes in seasonality of precipitation on the relative distribution of woody plants and grasses.

Effects of soil moisture and temperature on competitive interactions, particularly those involving different life-forms, are largely unknown. These topics are related to precipitation seasonality and edaphic features and can be investigated with experiments in controlled-environment chambers, greenhouses, and field environments. Relationships between woody and herbaceous plants should be studied at life-history stages ranging from seeds to mature plants.

The importance of short-term climatic events in the development of community structure represents a critical gap in the knowledge of ecosystems. Do periodic cold or hot temperatures constrain woody plant establishment? How do they affect different grasses (e.g., annual vs. perennial, native vs. introduced)? What role do extreme temperatures play in the mortality of established woody plants and grasses? What is the role of drought in shaping observed patterns of woody plant establishment? Recent research in other systems (Bassow et al. 1994) illustrates the kind of experiments that can significantly improve our understanding of short-term climatic controls on savanna plant communities.

Edaphic Factors

Soils and vegetation are inextricably linked, and soil properties are strongly correlated with overlying vegetation. Range site guides (see chapter 6) provide numerous detailed examples of the high correlation between soils and vegetation; these guides are widely available and offer a level of resolution appropriate to management. In some cases, these relationships may need to be refined. This is primarily a mapping exercise, which may be enhanced by geographical information systems technology.

Additional research in this area should be focused on determination of edaphic and geomorphic constraints on woody plant establishment. It has been suggested that argillic soil horizons constrain establishment of woody plants in southwestern grasslands and savannas (e.g., Loomis 1989, Archer 1994, McAuliffe 1994). McAuliffe (1994) hypothesized that water-impermeable argillic horizons reduce water availability to woody plants in summer below thresholds necessary for survival, or result in perched water tables in winter. Either phenomenon may contribute to woody plant mortality. Although these hypotheses are intuitively palatable and are often cited, they have not been tested.

Different types of soil may constrain recruitment of either woody plants or grasses. What is the importance of soil depth, particle size distribution, substrate (e.g., limestone vs. rhyolite), and nutrient content on vegetation distribution? Because soil properties are often highly correlated (e.g., shallow soils on steep slopes may be coarse textured and low in nitrogen and organic carbon), controlled experiments will be required to determine the relative importance of driving variables. Finally, fundamental information about the spatial and temporal distribution of soil nutrients, and the processes that affect these phenomena, is generally lacking for North American savannas.

Sociopolitical Issues

The role of sociopolitical factors on savannas is largely unknown. Much of the difficulty with this type of research stems from the fact that it is necessarily descriptive: legal and moral constraints preclude manipulative experiments on humans.

Effects of human actions on savannas at the landscape scale are essentially unexplored. As human densities and behaviors change, understanding how land tenure and changes in land use influence landscape-level processes will become increasingly important. A comparison of savannas characterized by different land ownerships may be a useful starting point for this research. For example, Californian oak and longleaf pine savannas occur primarily on private lands, whereas southwestern oak and ponderosa pine savannas occur primarily on public lands; investigations of differences in historical and contemporary development may yield important insights and allow prediction of future land-use patterns.

Little is known about the relative value humans place on savannas and savanna resources. In this regard, economic analyses can be used to describe the value of marketable products, and various techniques are available for assessing nonconsumptive resources (Kula 1994).

Summary

An understanding of the mechanisms underlying changes in the distribution of plants in North American savannas is critical to predicting the response of plant communities to management, disturbance, and environmental change. Descriptive studies (e.g., historical analyses, repeat photography, correlation, "natural" experiments), although important for assessment of pattern, are not well suited for elucidation of driving variables. Rather, the development of appropriate experiments is recommended for determining effects of anthropogenic and environmental factors on savanna ecosystems. Experiments should be field-based and focused on representative species and should be conducted at spatial and temporal scales most appropriate to management. Use of controlled-environment chambers or greenhouses enhances experimental control but may sacrifice realism, compared to field experiments. Controlled-environment trials should thus be used in conjunction with experimental field trials. Well-designed manipulative experiments should enable managers to predict effects of management, disturbance, and changing atmospheric, climatic, and hydrologic conditions on specific plant communities. Additional research should assess the importance of sociopolitical factors at the landscape scale.

Appendix of Scientific and Common Plant Names

Conifers[1]

Abies concolor (Gord. & Glend.) Lindl.	white fir
A. grandis (Dougl.) Lindl.	grand fir
Calocedrus decurrens (Torr.) Florin	incense-cedar
Juniperus ashei (Buch.)	Ashe juniper
J. deppeana (Steud.)	alligator juniper
J. flaccida (Schl.)	drooping juniper
J. erythrocarpa Cory	red-berry juniper
J. monosperma (Engelm.) Sarg.	one-seed juniper
J. occidentalis Hook.	western juniper
J. osteosperma (Torr.) Little	Utah juniper
J. pinchotii (Sudw.)	Pinchot juniper
J. scopulorum Sarg.	Rocky Mountain juniper
J. virginiana L.	eastern redcedar
Pinus cembroides Zucc.	border pinyon
P. echinata Mill.	shortleaf pine
P. edulis Engelm.	two-leaf pinyon
P. elliottii Engelm.	slash pine
P. monophylla T. & F.	single-leaf pinyon
P. palustris Mill.	longleaf pine
P. ponderosa Laws.	ponderosa pine
P. ponderosa var. *ponderosa*	Pacific ponderosa pine
P. ponderosa var. *arizonica* (Engelm.) Shaw	Arizona ponderosa pine
P. ponderosa var. *scopulorum* Engelm.	Rocky Mountain ponderosa pine
P. sabiniana Dougl.	foothill pine

[1] All conifers listed are capable of attaining tree size; many individuals are shrubs on some sites.

P. taeda L.	loblolly pine
Pseudotsuga menziesii Franco	Douglas-fir
Thuja plicata Donn	western redcedar

Angiosperms

Trees[2]

Acacia Mill.	acacia
Arbutus menziesii Pursh	Pacific madrone
Carya texana Buckl.	black hickory
Celtis laevigata Willd.	sugar hackberry
Cercocarpus ledifolius Nutt.	curlleaf mountain mahogany
Diospyros virginiana L.	common persimmon
Ilex L.	holly
Lithocarpus densiflorus (H. & A.) Rehd.	tanoak
Myrica cerifera L.	southern waxmyrtle
Prosopis L.	mesquite
P. glandulosa Torr.	honey mesquite
P. velutina Wooten	velvet mesquite
Quercus alba L.	white oak
Q. albocincta Trel.	cusi
Q. agrifolia Née	coast live oak
Q. arizonica Sarg.	Arizona white oak
Q. chihuahensis Trel.	Chihuahua oak
Q. chuchuichupensis C.H. Mueller	—
Q. douglasii H. & A.	blue oak
Q. ellipsoidalis E.J. Hill	northern pin oak
Q. emoryi Torr.	Emory oak
Q. engelmannii Greene	Engelmann oak
Q. garryana Dougl.	Oregon oak
Q. grisea Liebm.	gray oak
Q. hypoleucoides A. Camus	silverleaf oak
Q. incana Bartr.	bluejack oak
Q. laevis Walt.	turkey oak
Q. lobata Née	valley oak
Q. macrocarpa Michx.	bur oak
Q. marilandica Muenchh.	blackjack oak

[2]All taxa listed are capable of attaining tree size; some individuals are shrubs on some sites.

Q. margaretta Ashe	sandhill post oak
Q. oblongifolia Torr.	Mexican blue oak
Q. santaclarensis C.H. Mueller	—
Q. stellata Wang.	post oak
Q. velutina Lam.	black oak
Q. wislizenii A. DC.	interior live oak
Ulmus crassifolia Nutt.	cedar elm

Shrubs and Succulents

Artemisia L.	sagebrush
Capsicum annuum L. var. *aviculare* (Dierb.) D'Arcy & Eshb.	chiltepine
Flourensia cernua DC.	tarbush
Gutierrezia sarothrae (Pursh) Britton & Rusby	snakeweed
Larrea tridentata (DC.) Cov.	creosote bush
Nolina microcarpa S. Wats.	beargrass
Opuntia Mill.	cholla, prickly pear
Yucca L.	yucca
Y. elata Engelm.	soaptree yucca

Vines

Parthenocissus quinquefolia (L.) Planchon	Virginia creeper
Pueraria lobata (Willd.) Ohwi	kudzu
Rhus radicans L.	poison ivy

Graminoids[3]

Agropyron Gaertn.	wheatgrass
A. cristatum (L.) Gaertn.	crested wheatgrass
Andropogon L.	bluestem
A. gerardi Vitman	big bluestem
Aristida L.	three-awns
A. beyrichiana Trin. & Rupr.	wiregrass
A. stricta Michx.	wiregrass
Avena L.	wild oats

[3] All graminoids listed are members of the grass family (Poaceae) except sedges *(Cyperus, Carex)*, which are members of the Cyperaceae.

Bothriochloa ischaemum (L.) Keng var. *songarica* (Rupr.) Celarier & Harlan	King Ranch bluestem
Bouteloua Lag.	grama
B. curtipendula (Michx.) Torr.	sideoats grama
B. gracilis (H.B.K.) Lag. ex Stend.	blue grama
B. hirsuta Lag.	hairy grama
B. radicosa (Fourn.) Griffiths	purple grama
Bromus L.	brome
B. tectorum L.	downy brome
Buchlöe dactyloides (Nutt.) Engelm.	buffalograss
Carex L.	sedge
Cynodon dactylon (L.) Pers.	bermudagrass
Cyperus L.	sedge
Digitaria californica (Benth.) Henr.	Arizona cottontop
Elymus L.	rye
E. elymoides Swezey	squirreltail
Eragrostis Beauv.	lovegrass
E. curvula (Schrader) Nees.	weeping lovegrass
E. intermedia Hitchc.	plains lovegrass
E. lehmanniana Nees.	Lehmann lovegrass
E. mexicana Vasey	Mexican lovegrass
Festuca L.	fescue
Hilaria belangeri (Steud.) Nash	curly-mesquite
H. mutica (Buckl.) Benth.	tobosa
Hordeum L.	barley
Imperata cylindrica (L.) Beauv.	cogongrass
Lolium L.	rye
Muhlenbergia Schreber	muhly
M. emersleyi Vasey	bullgrass
M. longiligula Hitchc.	long-tongue muhly
Paspalum plicatulum Michx.	brownseed paspalum
Pennisetum ciliare (L.) Link.	buffelgrass
Poa pratensis L.	Kentucky bluegrass
Schizachyrium Nees.	bluestem
S. scoparium (Michx.) Nash	little bluestem
Sorghastrum nutans (L.) Nash	Indiangrass
Sorghum halepense (L.) Pers.	Johnsongrass
Sporobolus R. Br.	dropseed
Stipa L.	needlegrass
Taeniatherum asperum (Simonkai) Nevski	medusahead

Herbaceous Dicots

Erodium L'Her.	filaree
Euphorbia esula L.	leafy spurge
Geranium L.	geranium
Trifolium L.	clover

Note that several common names should be hyphenated because they represent taxonomic misnomers (e.g., eastern red-cedar is not a member of the cedar *[Cedrus]* genus). However, hyphenation is not used herein for species with widely accepted nonhyphenated common names.

Literature Cited

Abrahamson, W. G. 1984. Post-fire recovery of Florida Lake Wales Ridge vegetation. American Journal of Botany 71:9–21.

Ackerly, D. D., and Bazzaz, F. A. 1995. Plant growth and reproduction along CO_2 gradients: Non-linear responses and implications for community change. Global Change Biology 1:199–207.

Adams, D. M., Alig, R. J., McCarl, B. A., Callaway, J. M., and Winnett, S. M. 1996. An analysis of the impacts of public timber harvest policies on private forest management in the United States. Forest Science 42: 343–358.

Adams, T. E. Jr., Sands, P. B., Weitkamp, W. H., and McDougald, N. K. 1992. Oak seedling establishment on California rangelands. Journal of Range Management 45:93–98.

Adams, T. E. Jr., Sands, P. B., Weitkamp, W. H., McDougald, N. K., and Bartolome, J. 1987. Enemies of white oak regeneration in California. Pp. 459–462 in Plumb and Pillsbury 1987.

Aguirre, L., and Johnson, D. A. 1991. Influence of temperature and cheatgrass competition on seedling development of two bunchgrasses. Journal of Range Management 44:347–354.

Aldon, E. F., and Shaw, D. W. (tech. coords.). 1993. Managing Piñon-Juniper Ecosystems for Sustainability and Social Needs. USDA Forest Service Rocky Mountain Research Station General Technical Report RM-236. Fort Collins, Colo.

Alexander, R. R. 1986. Silvicultural Systems and Cutting Methods for Ponderosa Pine Forests in the Front Range of the Central Rocky Mountains. USDA Forest Service Rocky Mountain Research Station General Technical Report RM-128. Fort Collins, Colo.

Allen, B. H., Evett, R. R., Holzman, B. A., and Martin, A. J. 1989. Rangeland Cover Type Descriptions for California Hardwood Rangelands. Forest and Rangeland Resources Assessment Program, California Department of Forestry and Fire Protection, Sacramento.

Allen, C. D. (ed.). 1996. Fire Effects in Southwestern Forests: Proceedings of the Second La Mesa Fire Symposium. USDA Forest Service Rocky

Mountain Research Station General Technical Report RM-286. Fort Collins, Colo.

Allen, L. S. 1989. Livestock and the Coronado National Forest. Rangelands 11:9–13.

Allen, L. S. 1995. Fire management in the sky islands. Pp. 386–388 in DeBano et al. 1995.

Allen, T. F. H., O'Neill, R. V., and Hoekstra, T. W. 1984. Interlevel Relations in Ecological Research and Management: Some Working Principles from Hierarchy Theory. USDA Forest Service Rocky Mountain Research Station General Technical Report RM-110. Fort Collins, Colo.

Allen-Diaz, B. H., and Holzman, B. A. 1991. Blue oak communities in California. Madroño 38:80–95.

Allen-Diaz, B. H., Bartolome, J. W., and McClaran, M. P. 1998. Californian oak savanna. In Anderson et al. 1998. In press.

Allworth-Ewalt, N. A. 1982. Ornamental landscaping as a market for mesquite trees. Pp. P1–P7 in Parker 1982.

Amos, B. B., and Gehlbach, F. R. (eds.). 1988. Edwards Plateau Vegetation. Baylor University Press, Waco, Texas.

Anderson, R. C., Fralish, J. S., and Baskin, J. M. (eds.). 1998. Savanna, Barren, and Rock Outcrop Plant Communities of North America. Cambridge University Press, New York. In press.

Anderson, R. C., Schmidt, D., Anderson, M. R., and Gustafson, D. 1994. Parklands savanna restoration. Pp. 275–278 in Fralish et al. 1994.

Antunez de Mayolo, K. 1991. Oaks and environmental education. Pp. 273–277 in Standiford 1991.

Archer, S. 1989. Have southern Texas savannas been converted to woodlands in recent history? American Naturalist 134:545–561.

Archer, S. 1990. Development and stability of grass/woody mosaics in a subtropical savanna parkland, Texas, U.S.A. Journal of Biogeography 17:453–462.

Archer, S. 1993. Vegetation dynamics in changing environments. Rangelands Journal 15:104–116.

Archer, S. 1994. Woody plant encroachment into southwestern grasslands and savannas: rates, patterns, and proximate causes. Pp. 13–68 in Vavra et al. 1994.

Archer, S. 1995a. Harry Stobbs Memorial Lecture, 1993: Herbivore mediation of grass-woody plant interactions. Tropical Grasslands 29:218–235.

Archer, S. 1995b. Tree-grass dynamics in a *Prosopis*-thornscrub savanna parkland: Reconstructing the past and predicting the future. Ecoscience 2:83–99.

Archer, S. A., and Smeins, F. E. 1991. Ecosystem-level processes. Pp. 109–139 in Heitschmidt and Stuth 1991.

Archer, S., Schimel, D. S., and Holland, E. A. 1995. Mechanisms of shrubland expansion: Land use, climate, or CO_2? Climatic Change 29:91–99.

Archer, S., Scifres, C., Bassham, C. R., and Maggio, R. 1988. Autogenic succession in a subtropical savanna: Conversion of grassland to thorn woodland. Ecological Monographs 58:111–127.

Armentrout, S. M., and Pieper, R. D. 1988. Plant distribution surrounding Rocky Mountain pinyon pine and one seed juniper in south-central New Mexico. Journal of Range Management 41:139–143.

Arno, S. F., Harrington, M. G., Fiedler, C. E., and Carlson, C. E. 1995. Restoring fire-dependent ponderosa pine forests in western Montana. Restoration and Management Notes 13:32–36.

Arnold, J. F. 1964. Zonation of understory vegetation around a juniper tree. Journal of Range Management 17:41–42.

Arrhenius, S. 1896. On the influence of carbonic acid in the air upon the temperature of the ground. London, Edinburgh, and Dublin Philosophical Magazine and Journal of Science (series 5) 41:237–276.

Auclair, A. N. 1976. Ecological factors in the development of intensive-management ecosystems in the midwestern United States. Ecology 57:431–444.

Axelrod, D. I. 1937. A Pliocene flora from the Mt. Eden beds, southern California. Carnegie Institute of Washington Publication 476:125–183.

Axelrod, D. I. 1950. Studies in Late Tertiary Paleobotany. Carnegie Institute of Washington Publication 590. Washington, D.C.

Axelrod, D. I. 1958. Evolution of the Madro-Tertiary geoflora. Botanical Review 24:433–509.

Axelrod, D. I. 1978. History of the coniferous forests, California and Nevada. University of California Publications in Botany 70:1–62.

Axelrod, D. I. 1979. Age and origin of the Sonoran Desert. California Academy of Sciences Occasional Paper 132:1–74.

Bahre, C. J. 1991. A Legacy of Change: Historic Human Impact on Vegetation in the Arizona Borderlands. University of Arizona Press, Tucson.

Bahre, C. J. 1995. Human impacts on the grasslands of southeastern Arizona. Pp. 230–264 in McClaran and Van Devender 1995.

Bahre, C. J., and Shelton, M. L. 1993. Historic vegetation change, mesquite increases, and climate in southeastern Arizona. Journal of Biogeography 20:489–504.

Baisan, C. H., and Swetnam, T. W. 1990. Fire history on a desert mountain range: Rincon Mountain Wilderness, Arizona, U.S.A. Canadian Journal of Forest Research 20:1559–1569.

Balling, R. C., Meyer, G. A., and Wells, S. G. 1992. Climate change in Yellowstone National Park: Is the drought-related risk of wildfire increasing? Climatic Change 22:35–45.

Barbour, M. G. 1987. Community ecology and distribution of California hardwood forests and woodlands. Pp. 18–25 in Plumb and Pillsbury 1987.

Barbour, M. G., and Billings, W. D. (eds.). 1988. North American Terrestrial Vegetation. Cambridge University Press, New York.

Barbour, M. G., and Major, J. (eds.). 1977. Terrestrial Vegetation of California. Wiley, New York.

Barger, R. L., and Ffolliott, P. F. 1972. Physical Characteristics and Utilization of Major Woodland Tree Species in Arizona. USDA Forest Service Rocky Mountain Research Station Research Paper RM-83. Fort Collins, Colo.

Barkman, J. J. 1988. Some reflections on plant architecture and its ecological implications. Pp. 1–7 in Werger et al. 1988.

Barnes, P. W., and Archer, S. 1996. Influence of an overstory tree *(Prosopis glandulosa)* on associated shrubs in a savanna parkland: Implications for patch dynamics. Oecologia 105:493–500.

Barrett, J. W. (ed.). 1995. Regional Silviculture of the United States. 3rd ed. Wiley, New York.

Barth, R. C. 1980. Influence of pinyon pine trees on soil chemical and physical properties. Soil Science Society of America Journal 44:112–114.

Bartolome, J. W. 1987. California grassland and oak savannah. Rangelands 9:122–125.

Bartolome, J. W., Klukkert, S. E., and Barry, W. J. 1986. Opal phytoliths as evidence for displacement of native Californian grassland. Madroño 33:217–222.

Bartram, W. 1791. The Travels of William Bartram. James and Johnson, Philadelphia.

Baskin, J. M., and Baskin, C. C. 1989. Physiology of dormancy and germination in relation to seedbank ecology. Pp. 53–66 in Leck et al. 1989.

Bassow, S. L., McConnaughay, K. D. M., and Bazzaz, F. A. 1994. The response of temperate tree seedlings grown in elevated CO_2 to extreme temperature events. Ecological Applications 4:593–603.

Baumgartner, D. M., and Lotan, J. E. (eds.). 1987. Ponderosa Pine: The Species and Its Management. Proceedings of a Symposium, 29 September–1 October 1987, Spokane, Washington.

Bazzaz, F. A. 1990. The response of natural ecosystems to the rising global CO_2 levels. Annual Review of Ecology and Systematics 21:167–196.

Beale, E. F. 1858. Wagon road from Fort Defiance to the Colorado River. Sen. Exec. Doc. 124, 35th cong., 1st sess. GPO, Washington, D.C.

Bedunah, D. J., and Sosebee, R. E. (eds.). 1995. Wildland Plants: Physiological Ecology and Development Morphology. Society for Range Management, Denver.

Belsky, A. J. 1996. Viewpoint. Western juniper expansion: Is it a threat to arid northwestern ecosystems? Journal of Range Management 49: 53–59.

Bennett, D. A. 1995. Fuelwood harvesting in the sky islands of southeastern Arizona. Pp. 519–523 in DeBano et al. 1995.

Benson, L., and Darrow, R. A. 1981. Trees and Shrubs of the Southwestern Deserts. 3rd ed. University of Arizona Press, Tucson.

Betancourt, J. L., Van Devender, T. R., and Martin, P. S. 1990. Packrat Middens: The Last 40,000 Years of Biotic Change. University of Arizona Press, Tucson.

Betancourt, J. L., Pierson, E. A., Aasen, K. A., Fairchild-Parks, J. A., and Dean, J. S. 1993. Influence of history and climate on New Mexico piñon-juniper woodlands. Pp. 42–62 in Aldon and Shaw 1993.

Bewley, J. D., and Black, M. 1982. Physiology and Biochemistry of Seeds, vol. 2, Viability, Dormancy, and Environmental Control. Springer-Verlag, New York.

Bigelow, J. M. 1856. General description of the botanical character of the country. Pp. 1–16 in Reports of Explorations and Surveys to Ascertain the Most Practicable and Economical Route for a Railroad from the Mississippi River to the Pacific Ocean. Vol. 4. U.S. War Department. Beverly Tucker, Washington, D.C.

Billeb, E. W. 1968. Mining Camp Days. Howell North Books, Berkeley, Calif.

Billings, W. D. 1994. Ecological impacts of cheatgrass and resultant fire on ecosystems in the western Great Basin. Pp. 22–30 in Monsen and Kitchen 1994.

Blatner, K. A., and Govett, R. L. 1987. Ponderosa pine lumber market. Pp. 7–9 in Baumgartner and Lotan 1987.

Bock, J. H., and Bock, C. E. 1984. Effect of fires on woody vegetation in the pine-grassland ecotone of the southern Black Hills. American Midland Naturalist 112:35–42.

Bogusch, E. R. 1951. Climatic limits affecting distribution of mesquite *(Prosopis juliflora)* in Texas. Texas Journal of Science 3:554–558.

Bolsinger, C. L. 1988. The Hardwoods of California's Timberlands, Woodlands, and Savannas. USDA Forest Service Pacific Northwest Research Station Research Bulletin PNW-148. Portland, Ore.

Bonner, F. T., and Vozzo, J. A. 1987. Seed Biology and Technology of *Quercus*. USDA Forest Service Southern Research Station General Technical Report SO-66. New Orleans.

Bonser, S. P., and Reader, R. J. 1995. Plant competition and herbivory in relation to vegetation biomass. Ecology 76:2176–2183.

Borchert, M. 1994. Blue oak woodland. Pg. 11 in Shiflet, T. N. (ed.), Rangeland Cover Types of the United States. Society for Range Management, Denver.

Borchert, M. I., Davis, F. W., Michaelsen, J., and Oyler, L. D. 1989. Interactions of factors affecting seedling recruitment of blue oak *(Quercus douglasii)* in California. Ecology 70:389–404.

Born, J. D., Tymcio, R. P., and Casey, O. E. 1992. Nevada Forest Resources. USDA Forest Service Intermountain Research Station Resource Bulletin INT-76. Ogden, Utah.

Botts, P., Haney, A., Holland, K., and Packard, S. (coord. eds.). 1994. Midwest Oak Ecosystems Recovery Plan. Unpublished draft on file at Environmental Protection Agency, Great Lakes National Program Office, Chicago.

Bourne, A. 1820. On the prairies and barrens of the West. American Journal of Science 2:30–34.

Boyer, W. D. 1989. Response of planted longleaf pine bare-root and container stock to site preparation and release: Fifth-year results. Pp. 165–168 in Miller 1989.

Boyer, W. D. 1993. Regenerating longleaf pine with natural seeding. Proceedings of the Tall Timbers Fire Ecology Conference 18:299–309.

Boyer, W. D., and White, J. B. 1990. Natural regeneration of longleaf pine. Pp. 94–113 in Farrar 1990.

Brady, W. W., Stromberg, M. R., Aldon, E. F., Bohnam, C. D., and Henry, S. H. 1989. Response of a semidesert grassland to 16 years of rest from grazing. Journal of Range Management 42:284–288.

Bragg, T. B., and Hulbert, L. C. 1976. Woody plant invasion of unburned Kansas bluestem prairie. Journal of Range Management 29:19–24.

Bray, J. R. 1960. The composition of savanna vegetation in Wisconsin. Ecology 41:721–732.

Bridges, E. L., and Orzell, S. L. 1989. Longleaf pine communities of the west Gulf coastal plain. Natural Areas Journal 9:246–263.

Brock, J. H., Haas, R. H., and Shaver, J. C. 1978. Zonation of herbaceous vegetation associated with honey mesquite in northcentral Texas. Proceedings of the International Rangeland Congress 1:187–189.

Brown, A. L. 1950. Shrub invasion of southern Arizona desert grasslands. Journal of Range Management 11:129–132.

Brown, D. E. (ed.). 1982. Biotic communities of the American Southwest—United States and Mexico. Desert Plants 4:1–342.

Brown, J. H., and Heske, E. J. 1990. Control of a desert-grassland transition by a keystone rodent guild. Science 250:1705–1707.

Brown, J. H., and McDonald, W. 1995. Livestock grazing and conservation on southwestern rangelands. Conservation Biology 9:1644–1647.

Brown, J. H., Reichman, O. J., and Davidson, D. W. 1979. Granivory in desert ecosystems. Annual Review of Ecology and Systematics 10: 201–227.

Brown, J. K., Mutch, R. W., Spoon, C. W., and Wakimoto, R. H. (tech.

coords.). 1995. Proceedings: Symposium on Fire in Wilderness and Park Management. USDA Forest Service Intermountain Research Station General Technical Report INT-320. Ogden, Utah.

Brown, J. R., and Archer, S. 1989. Woody plant invasion of grasslands: Establishment of honey mesquite *(Prosopis glandulosa* var. *glandulosa)* on sites differing in herbaceous biomass and grazing history. Oecologia 80:19–26.

Brown, J. R., and Archer, S. 1990. Water relations of a perennial grass and seedling vs. adult woody plants in a subtropical savanna, Texas. Oikos 57:366–374.

Brown, J. R., and MacLeod, N. D. 1996. Integrating ecology into natural resource management policy. Environmental Management 30:289–296.

Bruce, D. 1951. Fire, site, and longleaf height growth. Journal of Forestry 49:25–28.

Bruner, A. D., and Klebenow, D. A. 1979. Predicting Success of Prescribed Fires in Pinyon-Juniper Woodland in Nevada. USDA Forest Service Intermountain Research Station Research Paper INT-219. Ogden, Utah.

Bryant, E. 1848. What I Saw in California. D. Appleton and Company, New York.

Buffington, L. D., and Herbel, C. H. 1965. Vegetational changes on a semidesert grassland range from 1958 to 1963. Ecological Monographs 35:139–164.

Burkhardt, J., and Tisdale, E. W. 1969. Natural and successional status of western juniper vegetation in Idaho. Journal of Range Management 22:264–270.

Burkhardt, J., and Tisdale, E. W. 1976. Causes of juniper invasion in southwestern Idaho. Ecology 57:472–484.

Bush, J. K., and Van Auken, O. W. 1990. Growth and survival of *Prosopis glandulosa* seedlings associated with shade and herbaceous competition. Botanical Gazette 151:234–239.

Bush, J. K., and Van Auken, O. W. 1991. Importance of time of germination and soil depth on growth of *Prosopis glandulosa* (Leguminosae) seedlings in the presence of a C_4 grass. American Journal of Botany 78:1732–1739.

Bush, J. K., and Van Auken, O. W. 1995. Woody plant growth related to planting time and clipping of a C_4 grass. Ecology 76:1603–1609.

Caldwell, M. M. 1990. Water parasitism stemming from hydraulic lift: A quantitative test in the field. Israel Journal of Botany 39:395–402.

Callaway, R. M. 1992. Effect of shrubs on recruitment of *Quercus douglasii* and *Quercus lobata* in California. Ecology 73:2118–2128.

Callaway, R. M. 1995. Positive interactions among plants. Botanical Review 61:306–349.

Callaway, R. M., and Nadkarni, N. M. 1991. Seasonal patterns of nutrient deposition in a *Quercus douglasii* woodland in central California. Plant and Soil 137:209–222.

Callaway, R. M., Nadkarni, N. M., and Mahall, B. E. 1991. Facilitation and interference of *Quercus douglasii* on understory productivity in central California. Ecology 72:1484–1499.

Campbell, B. D., Grime, J. P., Mackey, J. M. L., and Jalili, A. 1991. The quest for a mechanistic understanding of resource competition in plant communities: The role of experiments. Functional Ecology 5: 241–253.

Campbell, T. E. 1971. Cottontail Rabbits Clip Young Longleaf Pine Seedlings. USDA Forest Service Southern Research Station Research Note SO-130. New Orleans.

Cape, J. N., Brown, A. H. F., Robertson, S. M. C., Howson, G., and Paterson, I. S. 1991. Interspecies comparisons of throughfall and stemflow at three sites in northern Britain. Forest Ecology and Management 46:165–177.

Caprio, A. C., and Zwolinski, M. J. 1992. Fire effects on two oak species, *Quercus emoryi* and *Q. oblongifolia,* in southeastern Arizona. Pp. 150–154 in Ffolliott et al. 1992.

Caprio, A. C., and Zwolinski, M. J. 1995. Fire and vegetation in a Madrean oak woodland, Santa Catalina Mountains, southeastern Arizona. Pp. 389–398 in DeBano et al. 1995.

Carpenter, S. R. 1996. Microcosm experiments have limited relevance for community and ecosytem ecology. Ecology 77:677–680.

Carpenter, S. R., Chisholm, S. W., Krebs, C. J., Schindler, D. W., and Wright, R. F. 1995. Ecosystem experiments. Science 269:324–327.

Carter, M. G. 1964. Effects of drouth on mesquite. Journal of Range Management 17:275–276.

Ceballos, G., Mellink, E., and Hanebury, L. R. 1993. Distribution and conservation status of prairie dogs *(Cynomys mexicanus* and *Cynomys ludovicianus)* in Mexico. Biological Conservation 63:105–112.

Chapman, H. H. 1932. Is the longleaf pine type a climax? Ecology 13: 328–334.

Chapman, H. H. 1936a. Effect of fire in preparation of seedbed for longleaf pine seedlings. Journal of Forestry 34:852–854.

Chapman, H. H. 1936b. Effect of ground cover on growth rate of longleaf pine seedlings. Journal of Forestry 34:535.

Christensen, N. L. 1988. Vegetation of the southeastern coastal plain. Pp. 317–363 in Barbour and Billings 1988.

Clark, J. S. 1990. Landscape interactions among nitrogen, species composition, and long-term fire frequency. Biogeochemistry 11:1–22.

Clary, W. P. 1971. Effects of Utah juniper removal on herbage yields from Springerville soils. Journal of Range Management 24:373–378.

Clary, W. P. 1975. Present and future multiple use demands on the pinyon-juniper type. Pp. 19–26 in The Pinyon-Juniper Ecosystem: A Symposium. College of Natural Resources, Utah State University, Logan.

Clary, W. P. 1987. Silvicultural systems for forage production in ponderosa pine forests. Pp. 185–191 in Baumgartner and Lotan 1987.

Clawson, W. J. (ed.). 1989. Landscape Ecology: Study of Mediterranean Grazed Ecosystems. Proceedings of the Man and the Biosphere Symposium, XVI International Grassland Congress, October 7, 1989, Nice, France.

Clements, F. E. 1916. Plant Succession: An Analysis of the Development of Vegetation. Carnegie Institute of Washington Publication 242. Washington, D.C.

Clewell, A. F. 1989. Natural history of wiregrass (*Aristida stricta* Michx., Gramineae). Natural Areas Journal 9:223–233.

Cohen, Y., and Pastor, J. 1991. The response of a forest model to serial correlations of global warming. Ecology 71:1161–1165.

Cole, J., Lovett, G., and Findlay, S. (eds.). 1991. Comparative Analyses of Ecosystems. Springer-Verlag, Berlin.

Collins, D. C., and Green, A. W. 1988. South Dakota's Timber Resources. USDA Forest Service Intermountain Research Station Resource Bulletin INT-56. Ogden, Utah.

Conner, R. C., and Green, A. W. 1988. Colorado's Woodland Resources on State and Private Land. USDA Forest Service Intermountain Research Station Resource Bulletin INT-50. Ogden, Utah.

Conner, R. C., Born, J. D., Green, A. W., and O'Brien, R. A. 1990. Forest Resources of Arizona. USDA Forest Service Intermountain Research Station Resource Bulletin INT-69. Ogden, Utah.

Cooke, R. U., and Reeves, R. W. 1976. Arroyos and Environmental Change in the American South-west. Clarendon Press, Oxford, U.K.

Cooper, C. F. 1960. Changes in vegetation, structure, growth of southwestern pine forests since white settlement. Ecological Monographs 30:129–164.

Cornejo-Oviedo, E. H., Gronski, S., and Felker, P. 1992. Mature mesquite *(Prosopis glandulosa* var. *glandulosa)* stand description and preliminary effects of understory removal and fertilization on growth. Journal of Arid Environments 22:339–351.

Cornejo-Oviedo, E. H., Meyer, J. M., and Felker, P. 1991. Thinning dense sapling stands of mesquite *(Prosopis glandulosa* var. *glandulosa)* to optimize timber production and pasture improvement. Forest Ecology and Management 46:189–200.

Cornelius, J. M., Kemp, P. R., Ludwig, J. A., and Cunningham, G. L. 1991. The distribution of vascular plant species and guilds in space and time along a desert gradient. Journal of Vegetation Science 2:59–72.

Coupland, R. T. (ed.). 1991. Natural Grasslands: Introduction and Western Hemisphere. Ecosystems of the World 8A. Elsevier, Amsterdam.

Covington, W. W., and Moore, M. M. 1992. Postsettlement changes in natural fire regimes: Implications for restoration of old growth ponderosa pine forests. Pp. 81–99 in Kaufmann et al. 1992.

Covington, W. W., and Moore, M. M. 1994a. Postsettlement changes in natural fire regimes and forest structure: Ecological restoration of old-growth ponderosa pine forests. Journal of Sustainable Forestry 2: 153–181.

Covington, W. W., and Moore, M. M. 1994b. Southwestern ponderosa forest structure: Changes since Euro-American settlement. Journal of Forestry 92:39–47.

Covington, W. W., Everett, R. L., Steele, R., Irwin, L. L., Daer, T. A., and Auclair, A. N. D. 1994. Historical and anticipated changes in forest ecosystems of the Inland West of the United States. Journal of Sustainable Forestry 2:13–63.

Cox, J. R., De Alba-Avila, A., Rice, R. W., and Cox, J. N. 1993. Biological and physical factors influencing *Acacia constricta* and *Prosopis velutina* establishment in the Sonoran Desert. Journal of Range Management 46:43–48.

Cox, J. R., Martin-R., M. H., Ibarra-F., F. A., Fourie, J. H., Rethman, N. F., and Wilcox, D. G. 1988. The influence of climate and soils on the distribution of four African grasses. Journal of Range Management 41: 127–139.

Creighton, J. L., Zutter, B. R., Glover, G. R., and Gjerstad, D. H. 1987. Planted pine growth and survival responses to herbaceous vegetation control, treatment duration, and herbicide application technique. Southern Journal of Applied Forestry 11:223–227.

Croker, T. C. Jr. 1979. Longleaf pine: The longleaf pine story. Journal of Forest History 23:32–43.

Currie, P. O. 1975. Grazing Management of Ponderosa Pine–Bunchgrass Ranges of the Central Rocky Mountains: The Status of Our Knowledge. USDA Forest Service Rocky Mountain Research Station Research Paper RM-159. Fort Collins, Colo.

Curtis, J. T. 1959. The Vegetation of Wisconsin: An Ordination of Plant Communities. University of Wisconsin Press, Madison.

Curtis, P. S., Drake, B. G., and Whigham, D. F. 1989. Nitrogen and carbon dynamics in C_3 and C_4 estuarine marsh plants grown under elevated CO_2 in situ. Oecologia 78:297–301.

Cutter, B. E., and Guyette, R. P. 1994. Fire frequency on an oak-hickory ridgetop in the Missouri Ozarks. American Midland Naturalist 132: 393–398.

Dahl, B. E. 1982. Mesquite as a rangeland plant. Pp. A1–A20 in Parker 1982.

Dahl, B. E., Sosebee, R. E., Goen, J. P., and Bromley, C. S. 1978. Will mesquite control with 2,4,5-T enhance grass production? Journal of Range Management 31:129–131.

D'Antonio, C. M., and Vitousek, P. M. 1992. Biological invasions by exotic grasses, the grass/fire cycle, and global change. Annual Review of Ecology and Systematics 23:63–87.

Davis, M. B. 1989. Lags in vegetation response to greenhouse warming. Climatic Change 15:75–82.

DeBano, L. F., Ffolliott, P. F., Ortega-Rubio, A., Gottfried, G. J., Hamre, R. H., and Edminster, C. B. (tech. coords.). 1995. Biodiversity and Management of the Madrean Archipelago: The Sky Islands of Southwestern United States and Northwestern Mexico. USDA Forest Service Rocky Mountain Research Station General Technical Report RM-264. Fort Collins, Colo.

Delco, S., Beyer, R., and Allen, F. 1993. U.S. market for imported pignoli nuts. Pp. 164–167 in Aldon and Shaw 1993.

Delcourt, P. A., and Delcourt, H. R. 1987. Long-Term Forest Dynamics of the Temperate Zone: A Case Study of Late-Quaternary Forests in Eastern North America. Springer-Verlag, New York.

DESIPA [Department for Economic and Social Information and Policy Analysis]. 1993. World Population Prospects: The 1992 Revision. United Nations, New York.

Dettinger, M. D., Ghil, M., and Keppenne, C. L. 1995. Interannual and interdecadal variability in United States surface-air temperatures, 1910–1987. Climatic Change 31:35–66.

Diamond, J. M. 1986. Overview: Laboratory experiments, field experiments, and natural experiments. Pp. 3–22 in Diamond and Case 1986.

Diamond, J. M., and Case, T. J. (eds.). 1986. Community Ecology. Harper and Row, New York.

Dick-Peddie, W. A. 1993. New Mexico Vegetation: Past, Present, and Future. University of New Mexico Press, Albuquerque.

Dobyns, A. F. 1981. From Fire to Flood: Historic Human Destruction of Sonoran Desert Riverine Oases. Ballena Press Anthropological Paper Number 20. Socorro, N.Mex.

Drake, B. G., and Leadley, W. J. 1991. Canopy photosynthesis of crops and native plant communities exposed to long-term elevated CO_2. Plant, Cell and Environment 14:853–860.

Duever, L. C. 1989. Research priorities for the conservation, management, and restoration of wiregrass ecosystems. Natural Areas Journal 9:214–218.

Dulbecco, R. 1995. Jonas Salk (1914–95). Science 376:216.

Dyksterhuis, E. J. 1948. The vegetation of the western Cross Timbers. Ecological Monographs 18:325–376.

Dyksterhuis, E. J. 1949. Condition and management of rangeland based on quantitative ecology. Journal of Range Management 2:104–115.

Dyksterhuis, E. J. 1957. The savannah concept and its use. Ecology 38: 435–442.

Eagleson, P. S., and Segarra, R. I. 1985. Water-limited equilibrium of savanna vegetation systems. Water Resources Research 21:1483-1493.

Ehleringer, J. R., and Cooper, T. A. 1988. Correlations between carbon isotope ratio and microhabitat in desert plants. Oecologia 76:562–566.

Ehleringer, J. R., Sage, R. F., Flanagan, L. B., and Pearcy, R. W. 1991. Climate change and the evolution of C_4 photosynthesis. Trends in Ecology and Evolution 6:95–99.

Ehrlich, A. H. 1995. Implications of population pressure on agriculture and ecosystems. Advances in Botanical Research 21:79–104.

Eiten, G. 1986. The use of the term "savanna." Tropical Ecology 27:10–23.

Eiten, G. 1992. How names are used for vegetation. Journal of Vegetation Science 3:419–424.

Elliott, R. R. 1973. History of Nevada. University of Nebraska Press, Lincoln.

Ellison, L. 1960. Influence of grazing on plant succession of rangelands. Botanical Review 27:1–78.

Emory, W. H. 1857. Report on the United States and Mexican Boundary Survey. 2 vols. Sen. Exec. Doc. 108, 34th cong., 1st sess. GPO, Washington, D.C.

Engle, D. M., Stritzke, J. F., and Claypool, P. L. 1987. Herbage standing crop around eastern red cedar trees. Journal of Range Management 40:237–239.

Engstrom, R. T. 1993. Characteristic mammals and birds of longleaf pine forests. Proceedings of the Tall Timbers Fire Ecology Conference 18: 127–138.

Esser, G. 1992. Implications of climate change for production and decomposition in grasslands and coniferous forests. Ecological Applications 2:47–54.

Evans, J. 1987. Animal damage and its control in ponderosa pine forests. Pp. 109–114 in Baumgartner and Lotan 1987.

Evans, J. W. 1990. Powerful Rockey: The Blue Mountains and the Oregon Trail, 1811–1883. Eastern Oregon State College, La Grande.

Evans, R. A. 1988. Management of Pinyon-Juniper Woodlands. USDA Forest Service Intermountain Research Station General Technical Report INT-249. Ogden, Utah.

Everett, R. L. (comp.). 1987. Proceedings—Pinyon-Juniper Conference. USDA Forest Service Intermountain Research Station General Technical Report INT-215. Ogden, Utah.

Everett, R. L., Sharrow, S. H., and Meeuwig, R. O. 1983. Pinyon-juniper woodland understory distribution patterns and species association. Bulletin of the Torrey Botanical Club 110:454–463.

Evett, R. R. 1994. Determining environmental realized niches for six oak species in California through direct gradient analysis and ecological response surface modeling. Ph.D. diss., University of California, Berkeley.

Fagerstrom, T. 1987. On theory, data, and mathematics in ecology. Oikos 50:258–261.

Fajer, E. D., Bowers, M. D., and Bazzaz, F. A. 1989. The effects of enriched carbon dioxide atmospheres on plant/insect herbivore interactions. Science 243:1198–1200.

Fajer, E. D., Bowers, M. D., and Bazzaz, F. A. 1991. The effects of enriched CO_2 atmospheres on the buckeye butterfly *Junonia coenia.* Ecology 72:751–754.

Farrar, J. F., and Williams, M. L. 1991. The effects of increased atmospheric carbon dioxide and temperature on carbon partitioning, source-sink relations, and respiration. Plant, Cell and Environment 14:819–830.

Farrar, R. M. Jr. 1993. Growth and yield in naturally regenerated longleaf pine stands. Proceedings of the Tall Timbers Fire Ecology Conference 18:311–335.

Farrar, R. M. Jr. (ed.). 1990. Proceedings of the Symposium on the Management of Longleaf Pine. USDA Forest Service Southern Research Station General Technical Report SO-75. New Orleans.

Felger, R. S., and Wilson, M. F. 1995. Northern Sierra Madre Occidental and its Apachian outliers: A neglected center of biodiversity. Pp. 36–59 in DeBano et al. 1995.

Felker, P., Meyer, J. M., and Gronski, S. J. 1990. Application of self-thinning in mesquite *(Prosopis glandulosa* var. *glandulosa)* to range management and lumber production. Forest Ecology and Management 31:225–232.

Ferguson, E. R., Lawson, E. R., Maple, W. R., and Mesavage, C. 1968. Managing Eastern Redcedar. USDA Forest Service Southern Research Station Research Paper SO-37. New Orleans.

Ferris-Kaan, R. (ed.). 1995. The Ecology of Woodland Creation. Wiley, New York.

Ffolliott, P. F., Gottfried, G. J., Bennett, D. A., Hernandez C., V. M., Ortega-Rubio, A., and Hamre, R. H. (tech. coords.). 1992. Proceedings of the Symposium on Ecology and Management of Oak and Associated Woodlands: Perspectives in the Southwestern United States and Northern Mexico. USDA Forest Service General Technical Report RM-218. Fort Collins, Colo.

Fischer, C. E., Meadors, C. H., Behrens, R., Robinson, E. D., Marion, P. T., and Morton, H. L. 1959. Control of Mesquite on Grazing Lands. Bulletin 935. Texas Agricultural Experiment Station, College Station.

Fisher, R. F., Jenkins, M. J., and Fisher, J. W. 1987. Fire and the prairie-forest mosaic of Devil's Tower National Monument. American Midland Naturalist 117:250–257.

Flanagan, L. B., Ehleringer, J. R., and Marshall, J. D. 1992. Differential uptake of summer precipitation among co-occurring trees and shrubs in a pinyon-juniper woodland. Plant, Cell and Environment 15:831–836.

Flinn, R. C., Archer, S., Boutton, T. W., and Harlan, T. 1994. Identification of annual rings in an arid land woody plant, *Prosopis glandulosa.* Ecology 75:850–853.

Fortmann, L., and Huntsinger, L. 1987. Managing California's oak woodlands: A sociological study of owners. Pp. 379–384 in Plumb and Pillsbury 1987.

Fowells, H. A. (comp.). 1965. Silvics of Forest Trees of the United States. Agriculture Handbook 271. USDA Forest Service, Washington, D.C.

Fowler, W. P., and Ffolliott, P. F. 1995. A growth and yield model of Emory oak: Applications on watershed lands in southwestern United States. Pp. 347–350 in DeBano et al. 1995.

Fralish, J. S., Anderson, R. C., Ebinger, J. E., and Szafoni, R. (eds.). 1994. Proceedings of the North American Conference on Barrens and Savannas. Environmental Protection Agency, Great Lakes National Program Office, Chicago.

Franco-Pizaña, J., Fulbright, T. E., and Gardiner, D. T. 1995. Spatial relations between shrubs and *Prosopis glandulosa* canopies. Journal of Vegetation Science 6:73–78.

Franklin, J. F., and Dyrness, C. T. 1988. Natural Vegetation of Oregon and Washington. Oregon State University Press, Corvallis.

Fritts, H. C. 1976. Tree Rings and Climate. Academic Press, New York.

Frost, C. C. 1993. Four centuries of changing landscape patterns in the longleaf pine ecosystem. Proceedings of the Tall Timbers Fire Ecology Conference 18:17–43.

Frost, C. C., Walker, J., and Peet, R. K. 1986. Fire-dependent savannas and prairies of the southeast: Original extent, preservation status, and management problems. Pp. 248–357 in Kulhavy and Conner 1986.

Frost, W. E., and Edinger, S. B. 1991. Effects of tree canopies on soil characteristics of annual rangeland. Journal of Range Management 44: 286–288.

Frost, W. E., and McDougald, N. K. 1989. Tree canopy effects on herbaceous production of annual rangeland during drought. Journal of Range Management 42:281–283.

FRRAP [Forest and Rangeland Resources Assessment Program]. 1988. California's Forests and Rangelands: Growing Conflict over Changing Uses. California Department of Forestry and Fire Protection, Sacramento.

Fulbright, T. E., Kuti, J. O., and Tipton, A. R. 1995. Effects of nurse-plant canopy temperatures on shrub seed germination and seedling growth. Acta Oecologica 16:621–632.

Fulé, P. Z., and Covington, W. W. 1995. Changes in fire regimes and forest structures of unharvested Petran and Madrean pine forests. Pp. 408–415 in DeBano et al. 1995.

Garcia-Moya, E., and McKell, C. M. 1970. Contribution of shrubs to the nitrogen economy of a desert-wash plant community. Ecology 51: 81–88.

Gardner, J. L. 1951. Vegetation of the creosote bush area of the Rio Grande Valley in New Mexico. Ecological Monographs 21:379–403.

Garin, G. I. 1958. Longleaf pines can form vigorous sprouts. Journal of Forestry 56:430–431.

Gehlbach, R. R. 1981. Mountain Islands and Desert Seas. Texas A & M Press, College Station.

Gentry, H. S. 1946. Sierra Tauicharmona—a Sinaloa plant locale. Bulletin of the Torrey Botanical Club 73:356–362.

Gentry, H. S. 1957. Los pastizales de Durango: estudio ecologico, fisiografico, y floristico. Ediciones del Instituto Mexicano de Recursos Naturales Renovables, Mexico City.

Germaine, H. L. 1997. Constraints on establishment of *Quercus emoryi* at lower treeline. M.S. thesis, University of Arizona, Tucson.

Gersper, R. L., and Holowaychuk, N. 1971. Some effects of stemflow from forest canopy trees on chemical properties of soils. Ecology 52:691–702.

Ghil, M., and Vautgard, R. 1991. Interdecadal oscillations and the warming trend in global temperature time series. Nature 350:324–327.

Gilliam, F. S., Yurish, B. M., and Goodwin, L. M. 1993. Community composition of an old growth longleaf pine forest: Relationship to soil texture. Bulletin of the Torrey Botanical Club 120:287–294.

Giusti, G. A., Schmidt, R. H., and Churches, K. R. 1991. Oak sustainability: A challenge through public education and outreach programs. Pp. 246–249 in Standiford 1991.

Glendening, G. E. 1952. Some quantitative data on the increase of mesquite and cactus on a desert grassland range in southern Arizona. Ecology 33:319–328.

Glendening, G. E., and Paulsen, H. A. Jr. 1955. Reproduction and Establishment of Velvet Mesquite as Related to Invasion of Semidesert Grasslands. USDA Technical Bulletin 1127. Washington, D.C.

Glitzenstein, J. S., Platt, W. J., and Streng, D. R. 1995. Effects of fire regime and habitat on tree dynamics in north Florida longleaf pine savannas. Ecological Monographs 65:441–476.

Gordon, D. R., and Rice, K. J. 1993. Competitive effects of grassland annuals on soil water and blue oak *(Quercus douglasii)* seedlings. Ecology 74:68–82.

Gordon, D. R., Welker, J. M., Menke, J. W., and Rice, K. J. 1989. Competition for soil water between annual plants and blue oak *(Quercus douglasii)* seedlings. Oecologia 79:533–541.

Grace, S. L., and Platt, W. J. 1995. Effects of adult tree density and fire on the demography of pregrass stage juvenile longleaf pine (*Pinus palustris* Mill.). Journal of Ecology 83:75–86.

Gresham, C. A., Williams, T. M., and Lipscomb, D. J. 1991. Hurricane Hugo wind damage to southeastern U.S. coastal forest tree species. Biotropica 23:420–426.

Griffin, J. R. 1971. Oak regeneration in the upper Carmel Valley, California. Ecology 52:862–868.

Griffin, J. R. 1976. Regeneration in *Quercus lobata* savannas, Santa Lucia Mountains, California. American Midland Naturalist 95:422–435.

Griffin, J. R. 1977. Oak woodland. Pp. 383–415 in Barbour and Major 1977.

Grinnell, J. 1936. Up-hill planters. Condor 38:80–82.

Groisman, P. Ya., and Legates, D. R. 1995. Documenting and detecting long-term precipitation trends: Where we are and what should be done. Climatic Change 31:601–622.

Grubb, P. J. 1977. The maintenance of species-richness in plant communities: The importance of the regeneration niche. Biological Reviews 52:107–145.

Gruell, G. E. 1983. Fire and Vegetative Trends in the Northern Rockies: Interpretations from 1981–1982 Photographs. USDA Forest Service Intermountain Research Station General Technical Report INT-158. Ogden, Utah.

Guerin, D. N. 1993. Oak dome clonal structure and fire ecology in a Florida longleaf pine dominated community. Bulletin of the Torrey Botanical Club 120:107–114.

Gurevitch, J., and Collins, S. L. 1994. Experimental manipulation of natural plant communities. Trends in Ecology and Evolution 9:94–98.

Guyette, R. P., and Cutter, B. E. 1991. Tree-ring analysis of fire history of a post oak savanna in the Missouri Ozarks. Natural Areas Journal 11: 94–99.

Hadley, D., and Sheridan, T. E. 1995. Land Use History of the San Rafael Valley, Arizona (1540–1960). USDA Forest Service Rocky Mountain Research Station General Technical Report RM-269. Fort Collins, Colo.

Hairston, N. G. Sr. 1989. Ecological Experiments: Purpose, Design, and Execution. Cambridge University Press, Cambridge, U.K.

Hall, G., and Allen, R. 1980. Wood products from California oaks, Cal Oak Lumber Company style. Pp. 362–368 in Plumb 1980.

Hall, L. M., George, M. R., McCreary, D. D., and Adams, T. E. 1992. Effects of cattle grazing on blue oak seedling damage and survival. Journal of Range Management 45:503–506.

Harper, J. L. 1977. Population Biology of Plants. Academic Press, New York.

Harper, J. L. 1982. After description. Pp. 11–25 in Newman 1982.

Harper, J. L., Williams, J. T., and Sagar, G. R. 1965. The behaviour of seeds in soil: I. The heterogeneity of soil surfaces and its role in determining the establishment of plants from seed. Journal of Ecology 53: 273–286.

Harris, D. R. (ed.). 1980. Human Ecology in Savanna Environments. Academic Press, London.

Harris, G. A. 1967. Some competitive relationships between *Agropyron spicatum* and *Bromus tectorum*. Ecological Monographs 37:89–111.

Harris, G. A. 1977. Root phenology as a factor of competition among grass seedlings. Journal of Range Management 30:172–177.

Hastings, J. R., and Turner, R. M. 1965. The Changing Mile. University of Arizona Press, Tucson.

Havard, V. 1884. The mezquit. American Naturalist 18:451–459.

Hawkins, R. H. 1987. Applied hydrology in the pinyon-juniper zone. Pp. 493–504 in Everett 1987.

Haworth, K., and McPherson, G. R. 1994. Effects of *Quercus emoryi* on herbaceous vegetation in a semi-arid savanna. Vegetatio 112:153–159.

Haworth, K., and McPherson, G. R. 1995. Effects of *Quercus emoryi* trees on precipitation distribution and microclimate in a semi-arid savanna. Journal of Arid Environments 31:153–170.

Heikens, A. L., and Robertson, P. A. 1994. Barrens of the Midwest: A review of the literature. Castanea 59:184–194.

Heikens, A. L., and Robertson, P. A. 1995. Classification of barrens and other natural xeric forest openings in southern Illinois. Bulletin of the Torrey Botanical Club 122:203–214.

Heitschmidt, R. K., and Stuth, J. W. (eds.). 1991. Grazing Management: An Ecological Perspective. Timber Press, Portland, Ore.

Heitschmidt, R. K., Schultz, R. D., and Scifres, C. J. 1986. Herbaceous biomass dynamics and net primary production following chemical control of honey mesquite. Journal of Range Management 39:67–71.

Helvey, J. D., and Patric, J. H. 1965. Canopy and litter interception of rainfall by hardwoods of eastern United States. Water Resources Research 1:193–206.

Hendricks, D. M. 1985. Arizona Soils. College of Agriculture, University of Arizona, Tucson.

Hennessey, J. T., Gibbens, R. P., Tromble, J. M., and Cardenas, M. 1983. Vegetation changes from 1935 to 1980 in mesquite dunelands and former grasslands of southern New Mexico. Journal of Range Management 36:370–374.

Heyward, F. W. 1933. The root system of longleaf pine on the deep sands of western Florida. Ecology 14:136–148.

Heyward, F. W. 1939. The relation of fire to stand composition of longleaf pine forests. Ecology 20:287–304.

Hibbard, K. A., Archer, S., Valentine, D. W., and Schimel, D. S. 1993. Spatial variability in nitrogen mineralization in a *Prosopis* savanna parkland. Bulletin of the Ecological Society of America 74:275.

Hine, W. R. B. 1925. Hogs, fire, and disease versus longleaf pine. Southern Lumberman 119(1544):45–46.

Holland, V. L. 1973. A study of soil and vegetation under *Quercus douglasii* compared to open grassland. Ph.D. diss., University of California, Berkeley.

Holland, V. L. 1980. Effect of blue oak on rangeland forage production in central California. Pp. 314–318 in Plumb 1980.

Holland, V. L., and Morton, J. 1980. Effect of blue oak on nutritional quality of rangeland forage in central California. Pp. 319–322 in Plumb 1980.

Houghton, J. T., Jenkins, G. J., and Ephraums, J. J. (eds.). 1990. Climate Change: The IPCC Scientific Assessment. Published for the Intergovernmental Panel on Climate Change. Cambridge University Press, Cambridge, U.K.

Houghton, J. T., Callander, B. A., and Varney, S. K. (eds.). 1992. Climate Change 1992: the Supplementary Report to the IPCC Scientific Assessment. Published for the Intergovernmental Panel on Climate Change. Cambridge University Press, Cambridge, U.K.

Houghton, J. T., Miera, G., Filho, B., Callander, B. A., Harris, N., Kattenberg, A., and Maskell, K. (eds.). 1996. Climate Change 1995: the Science of Climate Change (The Second Assessment of Working Group I of the Intergovernmental Panel on Climate Change). Published for the Intergovernmental Panel on Climate Change. Cambridge University Press, Cambridge, U.K.

Hubbard, J. A., and McPherson, G. R. 1997. Acorn selection by Mexican jays: A test of a tri-trophic symbiotic relationship hypothesis. Oecologia 110:143–146.

Huber, D. W. 1992. Utilization of hardwoods, fuelwood, and special forest products in California, Arizona, and New Mexico. Pp. 103–108 in Ffolliott et al. 1992.

Huenneke, L. F., and Mooney, H. A. (eds.). 1989. Grassland Structure and Function: California Annual Grassland. Kluwer, Dordrecht.

Humphrey, R. R. 1958. The desert grassland. Botanical Review 24:193–253.

Humphrey, R. R. 1987. 90 Years and 535 Miles: Vegetation Changes along the Mexican Border. University of New Mexico Press, Albuquerque.

Huntley, B. J., and Walker, B. H. (eds.). 1982. Ecology of Tropical Savannas. Springer-Verlag, Berlin.
Huntly, N. 1991. Herbivores and the dynamics of communities and ecosystems. Annual Review of Ecology and Systematics 22:477–503.
Huntly, N., and Inouye, R. 1988. Pocket gophers in ecosystems: Patterns and mechanisms. Bioscience 38:786–793.
Huntsinger, L., and Clawson, W. J. 1989. The landscape of land use conflict in California. Pp. 133–151 in Clawson 1989.
Hutchison, M. D. 1994. The barrens of the Midwest: An historical perspective. Castanea 59:195–203.
Huxley, A. 1932. Brave New World. Chatto and Windus, London.
Idso, S. B. 1992. Shrubland expansion in the American Southwest. Climatic Change 22:85–86.
Idso, S. B., and Balling, R. C. 1992. United States drought trends of the past century. Agricultural and Forest Meteorology 60:279–284.
Idso, S. B., and Kimball, B. A. 1992. Seasonal fine-root biomass development of sour orange trees grown in atmospheres of ambient and elevated CO_2 concentration. Plant, Cell and Environment 15:337–341.
Idso, S. B., and Quinn, J. A. 1983. Vegetational Redistribution in Arizona and New Mexico in Response to a Doubling of the Atmospheric CO_2 Concentration. Climatological Publications; Science Paper Number 17. Laboratory of Climatology, Arizona State University, Tempe.
Idso, S. B., Kimball, B. A., and Hendrix, D. L. 1993. Air temperature modifies the size-enhancing effects of atmospheric CO_2 enrichment on sour orange tree leaves. Environmental and Experimental Botany 33: 293–299.
Jackson, L. E., Strauss, R. B., Firestone, M. K., and Bartolome, J. W. 1990. Influence of tree canopies on grassland productivity and nitrogen dynamics in deciduous oak savanna. Agriculture, Ecosystems and Environment 32:89–105.
James, F. J., Evans, J. D., Ralphs, M. H., and Child, R. D. (eds.). 1991. Noxious Range Weeds. Westview Press, Boulder, Colo.
Jameson, D. A. 1962. Effects of burning on a galleta-black grama range invaded by juniper. Ecology 43:760–763.
Jameson, D. A. 1966. Pinyon-juniper litter reduces growth of blue grama. Journal of Range Management 19:214–217.
Jameson, D. A. 1967. The relationship of tree overstory and herbaceous understory vegetation. Journal of Range Management 20:247–249.
Jameson, D. A. 1987. Climax or alternative steady states in woodland ecology. Pp. 9–13 in Everett 1987.
Jemison, G. M. 1943. Effect of litter removal on diameter growth of shortleaf pine. Journal of Forestry 41:213–214.
Johnsen, T. N. 1962. One-seed juniper invasion of northern Arizona grasslands. Ecological Monographs 32:187–207.

Johnson, C. G. Jr. 1994. Ponderosa pine-grassland. Pp. 8–9 in Shiflet 1994.

Johnson, E. A., and Gutsell, S. L. 1994. Fire frequency models, methods and interpretations. Advances in Ecological Research 25:239–287.

Johnson, F. L. 1986. Oak-hickory savannahs and transition zones: Preservation status and management problems. Pp. 345–347 in Kulhavy and Conner 1986.

Johnson, H. B., and Mayeux, H. S. 1990. *Prosopis glandulosa* and the nitrogen balance of rangelands: Extent and occurrence of nodulation. Oecologia 84:176–185.

Johnson, H. B., Polley, H. W., and Mayeux, H. S. 1993. Increasing CO_2 and plant-plant interactions: Effects on natural vegetation. Vegetatio 104/105:157–170.

Johnson, R. H., and Lincoln, D. E. 1991. Sagebrush carbon allocation patterns and grasshopper nutrition: The influence of carbon dioxide enrichment and soil mineral limitation. Oecologia 87:127–134.

Johnson, R. W., and Tothill, J. C. 1985. Definition and broad geographic outline of savanna lands. Pp. 1–13 in Tothill and Mott 1985.

Johnston, M. C. 1963. Past and present grassland of southern Texas and northeastern Mexico. Ecology 44:456–466.

Jones, P. D. 1993. Is climate change occurring? Evidence from the instrumental record. Pp. 27–44 in Kaiser and Drennen 1993.

Jones, P. D., Allen, L. H. Jr., and Jones, J. W. 1985. Responses of soybean canopy photosynthesis and transpiration to whole-day temperature changes in different CO_2 environments. Agronomy Journal 77:242–249.

Joyce, L. A. 1993. The life cycle of the range condition concept. Journal of Range Management 46:132–138.

Kaiser, H. M., and Drennen, T. E. (eds.). 1993. Agricultural Dimensions of Global Climate Change. St. Lucie Press, Delray Beach, Fla.

Kalisz, P. J., and Stone, E. L. 1984. The longleaf pine islands of the Ocala National Forest, Florida: A soil study. Ecology 65:1743–1754.

Kareiva, P. M., Kingsolver, J. G., and Huey, R. B. (eds.). 1993. Biotic Interactions and Global Change. Sinauer Associates, Sunderland, Mass.

Karl, T. R., Jones, P. D., Knight, R. W., Kukla, G., Razuvayev, V., Gallo, K. P., Lindseay, J., Charlson, R. J., and Peterson, T. C. 1993. A new perspective on recent global warming: Asymmetric trends of daily maximum and minimum temperature. Bulletin of the American Meteorological Society 74:1007–1023.

Katz, R. W., and Brown, B. G. 1992. Extreme events in a changing climate: Variability is more important than averages. Climatic Change 21:289–302.

Kaufmann, M. R., Moir, W. H., and Bassett, R. L. (eds.). 1992. Old Growth Forests of the Southwest and Rocky Mountain Regions: Proceedings of a Workshop. USDA Forest Service Rocky Mountain Research Station General Technical Report RM-213. Fort Collins, Colo.

Kay, B. L. 1987. Long-term effects of blue oak removal on forage production, forage quality, soil, and oak regeneration. Pp. 351–357 in Plumb and Pillsbury 1987.

Kay, B. L., and Leonard, O. A. 1980. Effect of blue oak removal on herbaceous forage production in the north Sierra foothills. Pp. 323–328 in Plumb 1980.

Keddy, P. A. 1989. Competition. Chapman and Hall, New York.

Kelly, J. F., and Bechtold, W. A. 1990. The longleaf pine resource. Pp. 11–22 in Farrar 1990.

Kessler, W. B., Salwasser, H., Cartwright, C. W. Jr., and Caplan, J. A. 1992. New perspectives for sustainable natural resources management. Ecological Applications 2:221–225.

Kimball, B. A., Rosenberg, N. J., Allen, L. H. Jr., Heichel, G. H., Stuber, C. W., Kissel, D. E., and Ernst, S. (eds.). 1990. Impact of Carbon Dioxide, Trace Gases, and Climate Change on Global Agriculture. ASA Special Publication Number 53. American Society of Agronomy, Crop Science Society of America, and Soil Science Society of America. Madison, Wisc.

Kimmins, J. P. 1987. Forest Ecology. Macmillan Publishing Co., New York.

Kingsolver, J. G., Huey, R. B., and Kareiva, P. M. 1993. An agenda for population and community research on global change. Pp. 480–486 in Kareiva et al. 1993.

Klemmedson, J. O., and Tiedemann, A. R. 1986. Long-term effects of mesquite removal on soil characteristics: II. Nutrient availability. Soil Science Society of America Journal 50:476–480.

Klopatek, J. M. 1987. Nutrient patterns and succession in pinyon-juniper ecosystems of northern Arizona. Pp. 391–396 in Everett 1987.

Klopatek, J. M., Olson, R. J., Emerson, C. J., and Jones, J. L. 1979. Land-use conflicts with natural vegetation in the United States. Environmental Conservation 6:191–200.

Knoop, W. T., and Walker, B. H. 1985. Interactions of woody and herbaceous vegetation in a southern African savanna. Journal of Ecology 73:235–253.

Knops, J. M. H., Nash, T. H. III, and Schlesinger, W. H. 1996. The influence of epiphytic lichens on the nutrient cycling of an oak woodland. Ecological Monographs 66:159–179.

Koch, G. W., and Mooney, H. A. (eds.). 1996. Carbon Dioxide and Terrestrial Ecosystems. Academic Press, San Diego.

Koniak, S., and Everett, R. L. 1982. Seed reserves in soils of successional stages in pinyon-juniper woodlands. American Midland Naturalist 108:295–303.

Koukoura, Z., and Menke, J. 1995. Competition for soil water between perennial bunch-grass (*Elymus glaucus* B.B.) and blue oak seedlings (*Quercus douglasii* H. & A.). Agroforestry Systems 32:225–235.
Krajina, V. J. 1965. Biogeoclimatic zones and biogeocoenoses of British Columbia. Ecology of Western North America 1:1–34.
Krammes, J. S. (tech. coord.). 1990. Effects of Fire Management of Southwestern Natural Resources. USDA Forest Service Rocky Mountain Research Station General Technical Report RM-191. Fort Collins, Colo.
Kula, E. 1994. Economics of Natural Resources, the Environment, and Policies. 2nd ed. Chapman and Hall, New York.
Kulhavy, D. L., and Conner, R. N. (eds.). 1986. Wilderness and Natural Areas in the Eastern United States: A Management Challenge. Stephen F. Austin State University Center for Applied Studies, Nacogdoches, Texas.
Laessle, A. M. 1958. The origin and successional relationship of sandhill vegetation and sand-pine scrub. Ecological Monographs 28:361–387.
Landers, J. L., and Mueller, B. S. 1985. Bobwhite Quail Management: A Habitat Approach. Miscellaneous Publication. Tall Timbers Research Station, Tallahassee, Fla.
Landers, J. L., Van Lear, D. H., and Boyer, W. D. 1995. The longleaf pine forests of the southeast: Requiem or renaissance? Journal of Forestry 93(11):39–44.
Larson, R. E., and Sodjoudee, M. E. 1982. Mesquite utilization: From the stump to finished product. Pp. O1–O12 in Parker 1982.
Lasueur, H. 1945. The Ecology of the Vegetation of Chihuahua, Mexico, North of Parallel Twenty-Eight. University of Texas Publication No. 4521. Austin.
Lawton, J. H. 1995. Ecological experiments with model systems. Science 269:328–331.
Leck, M. A., Parker, V. T., and Simpson, R. L. (eds.). 1989. Ecology of Soil Seedbanks. Academic Press, San Diego.
Lélé, S., and Norgaard, R. B. 1996. Sustainability and the scientist's burden. Conservation Biology 10:354–365.
Lemon, E. R. (ed.). 1983. CO_2 and Plants: The Response of Plants to Rising Levels of Atmospheric Carbon Dioxide. Westview Press, Boulder, Colo.
Lenhart, D. Y. 1934. Initial root development of longleaf pine. Journal of Forestry 32:459–461.
Leonard, S. G., Miles, R. L., and Summerfield, H. A. 1987. Soils of the pinyon-juniper woodlands. Pp. 227–230 in Everett 1987.
Leopold, A. 1924. Grass, brush, timber, and fire in southern Arizona. Journal of Forestry 22:1–10.
Levin, S. A. 1992. The problem of pattern and scale in ecology. Ecology 73:1943–1967.

Lewis, C. D., and Burgy, R. H. 1964. The relationship between oak tree roots and groundwater in fractured rock as determined by tritium tracing. Journal of Geophysical Research 69:2579–2588.

Lewis, H. T. 1973. Patterns of Indian Burning in California: Ecology and Ethnohistory. Ballena Press Anthropological Paper No. 1. Ramona, Calif.

Likens, G. E. 1985. An experimental approach for the study of ecosystems. Journal of Ecology 73:381–396.

Likens, G. E. (ed.). 1989. Long-Term Studies in Ecology. Springer-Verlag, Berlin.

Lincoln, D. E. 1993. The influence of plant carbon dioxide and nutrient supply on susceptibility to insect herbivores. Vegetatio 104/105:273–280.

Lincoln, D. E., and Couvet, D. 1989. The effect of carbon supply on allocation to allelochemicals and caterpillar consumption of peppermint. Oecologia 78:112–114.

Lincoln, D. E., Couvet, D., and Sionit, N. 1986. Response of an insect herbivore to host plants grown in enriched carbon dioxide atmospheres. Oecologia 69:556–560.

Lincoln, D. E., Fajer, E. D., and Johnson, R. H. 1993. Plant-insect herbivore interactions in elevated CO_2 environments. Trends in Ecology and Evolution 8:64–68.

Lindzen, R. 1993. Absence of scientific basis. National Geographic Research and Exploration 9:191–200.

Linhart, Y. B. 1987. Ecological and evolutionary studies of ponderosa pine in the Rocky Mountains. Pp. 77–89 in Baumgartner and Lotan 1987.

Little, E. L. Jr. 1941. Managing woodlands for piñon nuts. Chronica Botanica 6:348–349.

Little, E. L. Jr. 1993. Managing southwestern piñon-juniper woodlands: The past half century and the future. Pp. 105–107 in Aldon and Shaw 1993.

Livingston, M., Roundy, B. A., and Smith, S. E. 1995. Association of native grasses and overstory species in southern Arizona. Pp. 202–208 in Roundy et al. 1995.

Long, S. P. 1983. C_4 photosynthesis at low temperatures. Plant, Cell and Environment 6:345–363.

Long, S. P. 1991. Modification of the response of photosynthetic productivity to rising temperature by atmospheric CO_2 concentrations: Has its importance been underestimated? Plant, Cell and Environment 14:729–739.

Long, S. P., and Hutchin, P. R. 1991. Primary production in grasslands and coniferous forests with climate change: An overview. Ecological Applications 1:139–156.

Longhurst, W. M., Connolly, G. E., Browning, B. M., and Garton, E. O.

1979. Food interrelationships of deer and sheep in parts of Mendocino and Lake Counties. Hilgardia 47:191–249.

Longood, R., and Simmel, A. 1972. Organizational resistance to innovation suggested by research. Pp. 311–317 in Weiss 1972.

Loomis, L. E. 1989. Influence of heterogeneous subsoil development on vegetation patterns in a subtropical savanna parkland. Ph.D. diss., Texas A & M University, College Station.

Luken, J. O. 1990. Directing Ecological Succession. Chapman and Hall, New York.

MacCleery, D. W. 1995. The Way to a Healthy Future for National Forest Ecosystems in the West: What Role Can Silviculture and Prescribed Fire Play? USDA Forest Service Rocky Mountain Research Station General Technical Report RM-267. Fort Collins, Colo.

Madany, M. H., and West, N. E. 1980. Fire history of two montane forest areas of Zion National Park. Pp. 50–56 in Stokes and Dieterich 1980.

Madany, M. H., and West, N. E. 1983. Livestock grazing–fire regime interactions within montane forests of Zion National Park, Utah. Ecology 64:661–667.

Mahoney, M. J. 1976. Scientist as Subject: The Psychological Imperative. Ballinger Press, Cambridge, Mass.

Makus, D. J., Tiwari, S. C., Pearson, H. A., Haywood, J. D., and Tiarks, A. E. 1994. Okra production with pine straw mulch. Agroforestry Systems 27:121–127.

Manabe, S., and Wetherald, R. T. 1986. Reduction in summer soil wetness induced by an increase in atmospheric carbon dioxide. Science 232: 626–628.

Marañon, T., and Bartolome, J. W. 1993. Reciprocal transplants of herbaceous communities between *Quercus agrifolia* woodland and adjacent grassland. Journal of Ecology 81:673–682.

Marcy, R. B. 1849. Report of Captain R. B. Marcy. House Exec. Doc. 45, 31st cong., 1st sess. GPO, Washington, D.C.

Marshall, J. T. 1957. Birds of the pine-oak woodland in southern Arizona and adjacent Mexico. Pacific Coast Avifauna No. 32. Cooper Ornithological Society, Berkeley, Calif.

Martin, F. 1987. Effects of site, age, and density on growth of ponderosa pine in western Montana. Pp. 261–271 in Baumgartner and Lotan 1987.

Martin, P. S. 1975. Vanishings, and future of the prairie. Geoscience and Man 10:39–49.

Martin, S. C. 1975. Ecology and Management of Southwestern Semidesert Grass-Shrub Ranges: A Status of Knowledge. USDA Forest Service Rocky Mountain Research Station Research Paper RM-156. Fort Collins, Colo.

Martin, W. H., Boyce, S. G., and Echternacht, A. C. (eds.). 1993. Biodiversity of the Southeastern United States: Lowland Terrestrial Communities. Wiley, New York.

Matsuda, K., and McBride, J. R. 1989. Germination characteristics of selected California oak species. American Midland Naturalist 122: 66–76.

Matter, W. J., and Mannan, R. W. 1989. More on gaining reliable knowledge: A comment. Journal of Wildlife Management 53:1172–1176.

Mattoon, W. R. 1922. Longleaf Pine. USDA Bulletin 1061. GPO, Washington, D.C.

Mattson, W. 1980. Nitrogen and herbivory. Annual Review of Ecology and Systematics 11:119–162.

Mauchamp, A., and Janeau, J. L. 1993. Water funneling by the crown of *Flourensia cernua,* a Chihuahuan Desert shrub. Journal of Arid Environments 25:299–306.

Mayeux, H. S., Johnson, H. B., and Polley, H. W. 1991. Global change and vegetation dynamics. Pp. 62–74 in James et al. 1991.

Mayeux, H. S., Johnson, H. B., and Polley, H. W. 1994. Potential interactions between global change and intermountain annual grasslands. Pp. 95–100 in Monsen and Kitchen 1994.

McAuliffe, J. R. 1994. Landscape evolution, soil formation, and ecological patterns and processes in Sonoran Desert bajadas. Ecological Monographs 64:111–148.

McBride, J. R., Norberg, E., Cheng, S., and Mossadegh, A. 1991. Seedling establishment of coast live oak in relation to seed caching by jays. Pp. 143–148 in Standiford 1991.

McClain, W. E., Jenkins, M. A., Jenkins, S. E., and Ebinger, J. E. 1993. Changes in the woody vegetation of a bur oak savanna remnant in central Illinois. Natural Areas Journal 13:108–114.

McClaran, M. P., and Bartolome, J. W. 1989a. Effect of *Quercus douglasii* (Fagaceae) on herbaceous understory along a rainfall gradient. Madroño 36:141–153.

McClaran, M. P., and Bartolome, J. W. 1989b. Fire-related recruitment in stagnant *Quercus douglasii* populations. Canadian Journal of Forest Research 19:580–585.

McClaran, M. P., and McPherson, G. R. 1995. Can soil organic carbon isotopes be used to describe grass-tree dynamics at a savanna-grassland ecotone and within the savanna? Journal of Vegetation Science 6: 857–862.

McClaran, M. P., and McPherson, G. R. 1998. Oak savanna of the American Southwest. In Anderson et al. 1998. In press.

McClaran, M. P., and Van Devender, T. R. (eds.). 1995. The Desert Grassland. University of Arizona Press, Tucson.

McClaran, M. P., Allen, L. S., and Ruyle, G. B. 1992. Livestock production and grazing management in the encinal oak woodlands of Arizona. Pp. 57–64 in Ffolliott et al. 1992.

McDonald, B. 1995. The formation and history of the Malpai Borderlands Group. Pp. 483–486 in DeBano et al. 1995.

McDonald, P. M. 1980. Growth of thinned and unthinned hardwood stands in the northern Sierra Nevada . . . preliminary findings. Pp. 119–127 in Plumb 1980.

McLain, W. H. 1988. Arizona's 1984 Fuelwood Harvest. USDA Forest Service Intermountain Research Station Resource Bulletin INT-57. Ogden, Utah.

McLain, W. H. 1989. New Mexico's 1986 Fuelwood Harvest. USDA Forest Service Intermountain Research Station Resource Bulletin INT-60. Ogden, Utah.

McLeod, K. W., Sherrod, C. Jr., and Porch, T. E. 1979. Response of longleaf pine plantations to litter removal. Forest Ecology and Management 2:1–12.

McPherson, G. R. 1992a. Comparison of linear and non-linear overstory-understory models for ponderosa pine: A conceptual framework. Forest Ecology and Management 55:31–34.

McPherson, G. R. 1992b. Ecology of oak woodlands in Arizona. Pp. 23–33 in Ffolliott et al. 1992.

McPherson, G. R. 1993. Effects of herbivory and herbs on oak establishment in a semi-arid temperate savanna. Journal of Vegetation Science 4:687–692.

McPherson, G. R. 1994. Response of annual plants and communities to tilling in a semi-arid temperate savanna. Journal of Vegetation Science 5:415–420.

McPherson, G. R. 1995. The role of fire in desert grasslands. Pp. 130–151 in McClaran and Van Devender 1995.

McPherson, G. R., and Weltzin, J. F. 1997. Disturbance and Climate Change in U.S./Mexico Borderland Plant Communities: A State-of-the-Knowledge Review. USDA Forest Service Rocky Mountain Research Station General Technical Report. Fort Collins, Colo.

McPherson, G. R., and Wright, H. A. 1987. Factors affecting reproductive maturity of redberry juniper *(Juniperus pinchotii)*. Forest Ecology and Management 21:191–196.

McPherson, G. R., and Wright, H. A. 1989. Direct effects of competition on individual juniper plants: A field study. Journal of Applied Ecology 26:979–988.

McPherson, G. R., and Wright, H. A. 1990a. Effects of cattle grazing and *Juniperus pinchotii* canopy cover on herb cover and production in western Texas. American Midland Naturalist 123:144–151.

McPherson, G. R., and Wright, H. A. 1990b. Establishment of *Juniperus*

pinchotii in western Texas: Environmental effects. Journal of Arid Environments 19:283–287.
McPherson, G. R., Boutton, T. W., and Midwood, A. J. 1993. Stable carbon isotope analysis of soil organic matter illustrates vegetation change at the grassland/woodland boundary in southeastern Arizona, USA. Oecologia 93:95–101.
McPherson, G. R., Rasmussen, G. A., Wester, D. B., and Masters, R. A. 1991. Vegetation and soil zonation associated with *Juniperus pinchotii* Sudw. trees. Great Basin Naturalist 51:316–324.
McPherson, G. R., Wright, H. A., and Wester, D. B. 1988. Patterns of shrub invasion in semiarid Texas grasslands. American Midland Naturalist 120:391–397.
Medawar, P. 1984. Pluto's Republic. Oxford University Press, New York.
Medina, A. L. 1987. Woodland communities and soils of Fort Bayard, southwestern New Mexico. Journal of the Arizona-Nevada Academy of Sciences 21:99–112.
Melillo, J. M., McGuire, A. D., Kicklighter, D. W., Moore, B. III, Vorosmarty, C. J., and Schloss, A. L. 1993. Global climate change and terrestrial net primary production. Nature 363:234–240.
Menges, E. S., Abrahamson, W. G., Givens, K. T., Gallo, N. P., and Layne, J. N. 1993. Twenty years of vegetation change in five long unburned Florida plant communities. Journal of Vegetation Science 4:375–394.
Menke, J. W. 1989. Management controls on productivity. Pp. 173–199 in Huenneke and Mooney 1989.
Meyer, J. M., and Felker, P. 1990. Pruning mesquite *(Prosopis glandulosa* var. *glandulosa)* saplings for lumber production and ornamental use. Forest Ecology and Management 36:301–306.
Michler, N. Jr. 1850. Routes from the western boundary of Arkansas to Santa Fe and the valley of the Rio Grande. House Exec. Doc. 67, 31st cong., 1st sess. GPO, Washington, D.C.
Miller, B., Ceballos, G., and Reading, R. 1994. The prairie dog and biotic diversity. Conservation Biology 8:677–681.
Miller, B., Wemmer, C., Biggins, D., and Reading, R. 1990. A proposal to conserve black-footed ferrets and the prairie dog ecosystem. Environmental Management 14:763–769.
Miller, J. H. (comp.). 1989. Proceedings of the Fifth Biennial Southern Silvicultural Research Conference. USDA Forest Service Southern Research Station General Technical Report SO-74. New Orleans.
Miller, R. F., and Wigand, P. E. 1994. Holocene changes in semiarid pinyon-juniper woodlands. Bioscience 44:465–473.
Mitchell, J. F. B., Manabe, S., Meleshko, V., and Tokioka, T. 1990. Equilibrium climate change and its implications for the future. Pp. 131–172 in Houghton et al. 1990.

Mohr, C. 1901. Plant Life of Alabama. Brown Print Company, Montgomery, Ala.

Monsen, S. B., and Kitchen, S. G. (tech. coords.). 1994. Proceedings—Ecology and Management of Annual Rangelands. USDA Forest Intermountain Research Station General Technical Report INT-313. Ogden, Utah.

Mooney, H. A., and Chapin, F. S. III. 1994. Future directions of global change research in terrestrial ecosystems. Trends in Ecology and Evolution 9:371–372.

Mooney, H. A., Drake, B. G., Luxmoore, R. J., Oechel, W. C., and Pitelka, L. F. 1991. Predicting ecosystem responses to elevated CO_2 concentrations. Bioscience 41:96–104.

Morgan, P. 1987. Managing ponderosa pine forests for multiple objectives. Pp. 161–164 in Baumgartner and Lotan 1987.

Murphy, A. H., and Berry, L. J. 1973. Range pasture benefits through tree removal. California Agriculture 27:8–10.

Murphy, A. H., and Crampton, B. 1964. Quality of forage as affected by chemical removal of blue oak *(Quercus douglasii)*. Journal of Range Management 17:142–144.

Nabhan, G. P. 1985. Gathering the Desert. University of Arizona Press, Tucson.

Nasrallah, H. A., and Balling, R. C. 1993a. Analysis of recent climatic changes in the Arabian Gulf region. Environmental Conservation 20:223–226.

Nasrallah, H. A., and Balling, R. C. 1993b. Spatial and temporal analysis of Middle-Eastern temperature changes. Climatic Change 25:153–161.

Neftel, A., Moor, E., Oeschger, H., and Stauffer, B. 1985. Evidence from polar ice cores for the increase in atmospheric CO_2 in the past two centuries. Nature 315:45–57.

Neilson, R. P. 1986. High resolution climatic analysis and Southwest biogeography. Science 232:27–34.

Neilson, R. P. 1987. Biotic regionalization and climatic controls in western North America. Vegetatio 70:135–147.

Neilson, R. P. 1993. Transient ecotone response to climatic change: Some conceptual and modelling approaches. Ecological Applications 3:385–395.

Neilson, R. P. 1995. A model for predicting continental-scale vegetation distribution and water balance. Ecological Applications 5:362–385.

Neilson, R. P., and Marks, D. 1994. A global perspective of regional vegetation and hydrologic sensitivities from climatic change. Journal of Vegetation Science 5:715–730.

Neilson, R. P., and Wullstein, L. H. 1983. Biogeography of two southwest American oaks in relation to atmospheric dynamics. Journal of Biogeography 10:275–297.

Neilson, R. P., King, G. A., and Koerper, G. 1992. Toward a rule-based biome model. Landscape Ecology 7:27–32.

Nelson, L. R., Zutter, B. R., and Gjerstad, D. H. 1985. Planted longleaf pine seedlings respond to herbaceous weed control using herbicides. Southern Journal of Applied Forestry 9:236–240.

Nepstad, D. C., de Carvalho, C. R., Davidson, E. A., Jipp, P. H., Lefebvre, P. A., Negreiros, G. H., da Silva, E. D., Stone, T. A., Trumbore, S. E., and Vieira, S. 1994. The role of deep roots in the hydrological and carbon cycles of Amazonian forests and pastures. Nature 372:666–669.

Newman, E. I. (ed.). 1982. The Plant Community as a Working Mechanism. Blackwell, Oxford, U.K.

Norby, R. J., O'Neill, E. G., and Luxmoore, R. J. 1986. Effects of atmospheric CO_2 enrichment on the growth and mineral nutrition of *Quercus alba* seedlings in nutrient-poor soil. Plant Physiology 82:83–89.

Noss, R. F. 1989. Longleaf pine and wiregrass: Keystone components of an endangered ecosystem. Natural Areas Journal 9:211–213.

Nuzzo, V. A. 1986. Extent and status of midwest oak savanna: Presettlement and 1985. Natural Areas Journal 6:6–36.

Nyandiga, C. O., and McPherson, G. R. 1992. Germination of two warm-temperate oaks, *Quercus emoryi* and *Q. arizonica.* Canadian Journal of Forest Research 22:1395–1401.

Odum, E. P. 1989. Ecology and Our Endangered Life-Support Systems. 2nd ed. Sinauer Associates, Sunderland, Mass.

Oechel, W., and Strain, B. R. 1985. Native species responses to increased carbon dioxide concentration. Pp. 117–154 in Strain and Cure 1985.

Ortlieb, L., and Roldan-Q., J. (eds.). 1981. Geology of Northwestern Mexico and Southern Arizona. Estacion Regional del Noroeste, Instituto de Geologia, U.N.A.M., Hermosillo, Sonora.

Osbrink, W. L., Trumble, J. T., and Wagner, R. E. 1987. Host suitability of *Phaseolus lunata* for *Trichoplusia ni* (Lepidoptera:Noctuidae) in controlled carbon dioxide atmospheres. Environmental Entomology 16: 639–644.

Osmond, C. B., Pitelka, L. F., and Hidy, G. M. (eds.). 1990. Plant Biology of the Basin and Range. Ecological Studies 80. Springer-Verlag, New York.

OTA [Office of Technology Assessment]. 1993. Harmful Non-Indigenous Species in the United States. GPO, Washington, D.C.

Ovington, J. D., Heitkamp, D., and Lawrence, D. B. 1963. Plant biomass and productivity of prairie, savanna, oakwood, and maize field ecosystems in central Minnesota. Ecology 44:52–63.

Padien, D. J., and Lajtha, K. 1992. Plant spatial pattern and nutrient distribution in pinyon-juniper woodlands along an elevational gradient in northern New Mexico. International Journal of Plant Science 153: 425–433.

Parker, H. W. (ed.). 1982. Mesquite Utilization, 1982. Texas Tech Press, Lubbock.

Parker, K. W. 1954. Application of ecology in the determination of range condition and trend. Journal of Range Management 7:14–23.

Parker, V. T. 1977. Dominance relationships of tree associated herbs in some California grasslands. Ph.D. diss., University of California, Santa Barbara.

Parker, V. T. 1994. Coast live oak woodland. Pp. 12–13 in Shiflet 1994.

Parker, V. T., and Billow, C. R. 1987. Survey of soil nitrogen availability beneath evergreen and deciduous species of *Quercus*. Pp. 98–102 in Plumb and Pillsbury 1987.

Parker, V. T., and Muller, C. H. 1982. Vegetational and environmental changes beneath isolated live oak trees *(Quercus agrifolia)* in a California annual grassland. American Midland Naturalist 107:69–81.

Patterson, D. T., and Flint, E. P. 1990. Implications of increasing carbon dioxide and climate change for plant communities and competition in natural and managed ecosystems. Pp. 83–110 in Kimball et al. 1990.

Paulsen, H. A. Jr. 1950. Mortality of velvet mesquite seedlings. Journal of Range Management 3:281–286.

Pavlik, B. M., Muick, P. C., Johnson, S., and Popper, M. 1991. Oaks of California. Cachuma Press, Los Olivos, Calif.

Pearcy, R. W., and Bjorkman, O. 1983. Physiological effects. Pp. 65–105 in Lemon 1983.

Pearson, G. A. 1923. Natural Reproduction of Western Yellow Pine in the Southwest. USDA Department Bulletin 1105. GPO, Washington, D.C.

Pearson, G. A. 1942. Herbaceous vegetation a factor in natural regeneration of ponderosa pine in the Southwest. Ecological Monographs 12: 313–338.

Pearson, G. A. 1950. Management of Ponderosa Pine in the Southwest. USDA Forest Service Agriculture Monograph 6. GPO, Washington, D.C.

Pearson, G. A. 1951. A comparison of the climate in four ponderosa pine regions. Journal of Forestry 49:256–258.

Pearson, H. A., Grelen, H. E., Parresol, B. R., and Wright, V. L. 1987. Detailed vegetative description of the longleaf-slash pine type, Vernon District, Kisatchie National Forest, Louisiana. Pp. 107–115 in Pearson et al. (comps.) 1987.

Pearson, H. A., Smeins, F. E., and Thill, R. E. (comps.). 1987. Ecological, Physical, and Socioeconomic Relationships within Southern National Forests. USDA Forest Service Southern Research Station General Technical Report SO-68. New Orleans.

Peet, R. K. 1988. Forests of the Rocky Mountains. Pp. 63–101 in Barbour and Billings 1988.

Peet, R. K. 1993. A taxonomic study of *Aristida stricta* and *A. beyrichiana.* Rhodora 95:25–37.

Peet, R. K., and Allard, D. J. 1993. Longleaf pine vegetation of the southern Atlantic and eastern Gulf coast regions: A preliminary classification. Proceedings of the Tall Timbers Fire Ecology Conference 18: 45–81.

Pessin, L. J. 1938. The effect of vegetation on the growth of longleaf pine seedlings. Ecological Monographs 8:115–149.

Pessin, L. J. 1939. Density of stocking and character of ground cover a factor in longleaf pine reproduction. Journal of Forestry 37:255–258.

Pessin, L. J. 1944. Stimulating the early height growth of longleaf pine seedlings. Journal of Forestry 42:95–98.

Pessin, J. L., and Chapman, R. A. 1944. The effect of living grass on the growth of longleaf pine seedlings in pots. Ecology 25:85–90.

Peters, R. H. 1991. A Critique for Ecology. Cambridge University Press, Cambridge, U.K.

Pickett, S. T. A., Kolasa, J., and Jones, C. G. 1994. Ecological Understanding: The Nature of Theory and the Theory of Nature. Academic Press, San Diego.

Pickford, G. D. 1932. The influence of heavy grazing and of promiscuous burning on spring-fall ranges in Utah. Ecology 13:159–171.

Pillsbury, N. H., and Joseph, J. P. 1991. Coast live oak thinning study in the central coast of California—fifth-year results. Pp. 320–332 in Standiford 1991.

Pillsbury, N. H., DeLasaux, M. J., and Plumb, T. R. 1987. Coast live oak thinning study in the central coast of California. Pp. 92–97 in Plumb and Pillsbury 1987.

Pinchot, G. 1899. The relation of forests and forest trees. National Geographic 10:393–403.

Platt, J. R. 1964. Strong inference. Science 146:347–353.

Platt, W. J., Evans, G. W., and Davis, M. M. 1988a. Effects of fire season on flowering forbs and shrubs in longleaf pine forest. Oecologia 76: 353–363.

Platt, W. J., Evans, G. W., and Rathburn, S. L. 1988b. The population dynamics of a long-lived conifer *(Pinus palustris).* American Naturalist 131:491–525.

Platt, W. J., Glitzenstein, J. S., and Streng, D. S. 1991. Evaluating pyrogenicity and its effects on vegetation in longleaf pine savannas. Proceedings of the Tall Timbers Fire Ecology Conference 17:143–161.

Plumb, T. R. (ed.). 1980. Proceedings of the Symposium on the Ecology, Management, and Utilization of California Oaks. USDA Forest Ser-

vice Pacific Southwest Research Station General Technical Report PSW-44. Berkeley, Calif.
Plumb, T. R., and Pillsbury, N. H. (tech. coords.). 1987. Proceedings of the Symposium on Multiple-Use Management of California's Hardwood Resources. USDA Forest Service Pacific Southwest Research Station General Technical Report PSW-100. Berkeley, Calif.
Polley, H. W., Johnson, H. B., Marino, B. D., and Mayeux, H. S. 1993. Increase in C_3 plant water-use efficiency and biomass over glacial to present CO_2 concentrations. Nature 361:61–64.
Polley, H. W., Johnson, H. B., and Mayeux, H. S. 1992. Carbon dioxide and water fluxes of C_3 annuals and C_3 and C_4 perennials at subambient CO_2 concentrations. Functional Ecology 6:693–703.
Polley, H. W., Johnson, H. B., and Mayeux, H. S. 1994. Increasing CO_2: Comparative responses of the C_4 grass *Schizachyrium* and grassland invader *Prosopis*. Ecology 75:976–988.
Popper, K. 1981. Science, pseudo-science, and falsifiability. Pp. 92–99 in Tweney et al. 1981.
Potter, L. D., and Green, D. L. 1964. Ecology of ponderosa pine in western North Dakota. Ecology 45:10–23.
Potvin, C., and Strain, B. R. 1985. Effects of CO_2 enrichment and temperature on growth in two C4 weeds, *Echinochloa crusgalli* and *Eleusine indica*. Canadian Journal of Botany 63:1495–1499.
Prentice, I. C. 1986. Vegetation responses to past climatic variation. Vegetatio 67:131–141.
Price, P. W. 1991. The plant vigor hypothesis and herbivore attack. Oikos 62:244–251.
Price, P. W., Lewinsohn, T. M., Fernandes, G. W., and Benson, W. W. (eds.). 1991. Plant-Animal Interactions: Evolutionary Ecology in Tropical and Temperate Regions. Wiley, New York.
Progulske, D. R. 1974. Yellow Ore, Yellow Hair, Yellow Pine: A Photographic Study of a Century of Forest Ecology. Agricultural Experiment Station Bulletin 616. South Dakota State University, Brookings.
Pyne, S. J. 1984. Fire in America: A Cultural History of Wildland and Rural Fire. Princeton University Press, Princeton, N.J.
Quarles, S. L. 1987. Overview of the hardwood utilization problem. Pp. 233–236 in Plumb and Pillsbury 1987.
Ratliff, R. D., Duncan, D. A., and Westfall, S. E. 1991. California oak-woodland overstory species affect herbage understory: Management implications. Journal of Range Management 44:306–310.
Raven, P. H., and Axelrod, D. I. 1978. Origin and relationships of the California flora. University of California Publications in Botany 72: 1–134.
Reinke, J. J., Adriano, D. C., and McLeod, K. W. 1981. Effects of litter alter-

ation on carbon dioxide evolution from a South Carolina pine forest floor. Soil Science Society of America Journal 45:620–623.

Reynolds, H. G. 1958. The ecology of Merriam kangaroo rats (*Dipodomys merriami* Mearns) on grazing lands of southern Arizona. Ecological Monographs 28:111–127.

Reynolds, H. G., and Glendening, G. E. 1949. Merriam kangaroo rats a factor in mesquite propagation on southern Arizona rangelands. Journal of Range Management 2:193–197.

Reynolds, H. G., and Glendening, G. E. 1950. Relation of kangaroo rats to range vegetation in southern Arizona. Ecology 31:456–463.

Rice, E. L., and Penfound, W. T. 1959. The upland forests of Oklahoma. Ecology 40:593–608.

Rice, K. J. 1987. Interaction of disturbance patch size and herbivory in *Erodium* colonization. Ecology 68:1113–1115.

Rice, K. J., Gordon, D. R., Hardison, J. L., and Welker, J. M. 1993. Phenotypic variation in seedlings of a "keystone" tree species *(Quercus douglasii):* The interactive effects of acorn source and competitive environment. Oecologia 96:537–547.

Risser, P. G. (ed.). 1991. Long-Term Ecological Research: An International Perspective. Wiley, New York.

Roberts, P. R., and Oosting, H. J. 1958. Responses of venus fly trap *(Dionaea muscipula)* to factors involved in its endemism. Ecological Monographs 28:193–218.

Rogers, G. F. 1982. Then and Now: A Photographic History of Vegetation Change in the Central Great Basin Desert. University of Utah Press, Salt Lake City.

Rogers, G. F., and Vint, M. K. 1987. Winter precipitation and fire in the Sonoran Desert. Journal of Arid Environments 13:47–52.

Roise, J. P., Chung, J., and Lancia, R. 1991. Red-cockaded woodpecker habitat management and longleaf pine straw production: An economic analysis. Southern Journal of Applied Forestry 15:88–92.

Romesburg, H. C. 1981. Wildlife science: Gaining reliable knowledge. Journal of Wildlife Management 45:293–313.

Rosene, W. 1969. The Bobwhite Quail: Its Life and Management. Rutgers University Press, New Brunswick, N.J.

Rosson, J. F. Jr. 1994. *Quercus stellata* growth and stand characteristics in the *Quercus stella–Quercus marilandica* forest type in the Cross Timbers region of central Oklahoma. Pp. 329–331 in Fralish et al. 1994.

Roundy, B. A., and Jordan, G. L. 1988. Vegetation changes in relation to livestock exclusion and rootplowing in southeastern Arizona. Southwestern Naturalist 33:425–436.

Roundy, B. A., McArthur, E. D., Haley, J. S., and Mann, D. K. (comps.). 1995. Proceedings: Wildland Shrub and Arid Land Restoration Sym-

posium. USDA Forest Service Intermountain Research Station General Technical Report INT-315. Ogden, Utah.
Rowlands, P. G. 1993. Climatic factors and the distribution of woodland vegetation in the Southwest. Southwestern Naturalist 38:185–197.
Rozema, J., Lambers, H., van de Geijn, S. C., and Cambridge, M. L. (eds.). 1993. CO_2 and Biosphere. Kluwer Academic Publishers, Dordrecht.
Ruffin, E. 1843. The Private Diary of Edmund Ruffin. Edited by Mathew, W. M. University of Georgia Press, Athens.
Rummell, R. S. 1951. Some effects of livestock grazing on ponderosa pine forest and range in central Washington. Ecology 32:594–607.
Rzedowski, J. 1983. Vegetacion de Mexico. 2nd ed. Editorial Limusa, Mexico City.
Sala, O. E., Golluscio, R. A., Lauenroth, W. K., and Soriano, A. 1989. Resource partitioning between shrubs and grasses in the Patagonian steppe. Oecologia 81:501–505.
Sanchini, P. J. 1981. Population structure and fecundity patterns in *Quercus emoryi* and *Q. arizonica* in southeastern Arizona. Ph.D. diss., University of Colorado, Boulder.
San Jose, J. J., and Montes, R. 1992. Rainfall partitioning by a semideciduous forest grove in the savannas of the Orinoco Llano, Venezuela. Journal of Hydrology 132:249–262.
Sarmiento, G. 1984. The Ecology of Neotropical Savannas. Harvard University Press, Cambridge, Mass.
Savage, M. 1991. Structural dynamics of a southwestern pine forest under chronic human influence. Annals of the Association of American Geographers 81:271–289.
Savage, M., and Swetnam, T. W. 1990. Early 19th-century fire decline following sheep pasturing in a Navajo ponderosa pine forest. Ecology 71:2374–2378.
Scanlan, J. C., and Archer, S. 1991. Simulated dynamics of succession in a North American subtropical *Prosopis* savanna. Journal of Vegetation Science 2:625–634.
Scarnecchia, D. L. 1995. The rangeland condition concept and range science's search for identity: A systems viewpoint. Journal of Range Management 48:181–186.
Schafale, M. P., and Weakley, A. S. 1989. Ecological concerns about pine straw raking in southeastern longleaf pine ecosystems. Natural Areas Journal 10:220–221.
Schlesinger, W. H., Reynolds, J. F., Cunningham, G. L., Huenneke, L., Jarrell, W. M., Virginia, R. A., and Whitford, W. G. 1990. Biological feedbacks in global desertification. Science 247:1043–1048.
Schmidt, T. L., and Stubbendieck, J. 1993. Factors influencing eastern redcedar seedling survival on rangeland. Journal of Range Management 46:448–451.

Schmutz, E. M., Smith, E. L., Ogden, P. R., Cox, M. L., Klemmedson, J. O., Norris, J. J., and Fierro, L. C. 1991. Desert grassland. Pp. 337–362 in Coupland 1991.

Schott, M. R., and Pieper, R. D. 1985. Influence of canopy characteristics of one-seeded juniper on understory grasses. Journal of Range Management 38:328–331.

Schubert, G. H. 1974. Silviculture of Southwestern Ponderosa Pine: the Status of Our Knowledge. USDA Forest Service Rocky Mountain Research Station Research Paper RM-123. Fort Collins, Colo.

Schule, W. 1991. Human evolution, animal behavior, and Quaternary extinctions: A paleo-ecology of hunting. Homo Gottingen 41:228–250.

Schuster, J. L. 1964. Root development of native plants under three grazing intensities. Ecology 45:63–70.

Schwab, B. A. 1993. Bureau of Indian Affairs pilot woodlands management program. Pp. 146–148 in Aldon and Shaw 1993.

Schwarz, G. F. 1907. The Longleaf Pine in Virgin Forest: A Silvical Study. Wiley, New York.

Scifres, C. J., and Brock, J. H. 1969. Moisture-temperature interrelations in germination and early seedling development of mesquite. Journal of Range Management 22:334–337.

Scott, T. A. 1991. The distribution of Engelmann oak *(Quercus engelmannii)* in California. Pp. 351–359 in Standiford 1991.

Sharitz, R. R., Boring, L. R., Van Lear, D. H., Pinder, J. E. III. 1992. Integrating ecological concepts with natural resource management of southern forests. Ecological Applications 2:226–237.

Shiflet, T. N. (ed.). 1994. Rangeland Cover Types of the United States. Society for Range Management, Denver.

Shoulders, E. 1990. Identifying longleaf pine sites. Pp. 23–37 in Farrar 1990.

Shreve, F. 1939. Observations on the vegetation of Chihuahua. Madroño 5:1–13.

Simberloff, D. 1983. Competition theory, hypothesis-testing, and other community ecology buzzwords. American Naturalist 122:626–635.

Skarpe, C. 1992. Dynamics of savanna ecosystems. Journal of Vegetation Science 3:293–300.

Skovlin, J. M., and Thomas, J. W. 1995. Interpreting Long-Term Trends in Blue Mountain Ecosystems from Repeat Photography. USDA Forest Service Pacific Northwest Research Station General Technical Report PNW-315. Portland, Ore.

Slansky, F., and Rodriguez, J. G. (eds.). 1987. Nutritional Ecology of Insects, Mites, Spiders, and Related Invertebrates. Wiley, New York.

Smeins, F. E., and Diamond, D. D. 1986. Grasslands and savannahs of east central Texas: Ecology, preservation status, and management problems. Pp. 381–394 in Kulhavy and Conner 1986.

Smeins, F. E., and Merrill, L. B. 1988. Long-term change in semi-arid grassland. Pp. 101–114 in Amos and Gehlbach 1988.

Smeins, F. E., Taylor, T. W., and Merrill, L. B. 1976. Vegetation of a 25-year exclosure on the Edwards Plateau, Texas. Journal of Range Management 29:24–29.

Smith, D. A., and Schmutz, E. M. 1975. Vegetative changes on protected versus grazed desert grassland ranges in Arizona. Journal of Range Management 28:453–457.

Smith, D. M. 1986. The Practice of Silviculture. 8th ed. Wiley, New York.

Smith, M. A., Wright, H. A., and Schuster, J. L. 1975. Reproductive characteristics of redberry juniper. Journal of Range Management 35:126–128.

Smith, S. D., and Stubbendieck, J. 1990. Production of tall-grass prairie herbs below eastern redcedar. Prairie Naturalist 22:13–18.

Smith, S. D., Strain, B. R., and Sharkey, T. D. 1987. Effects of CO_2 enrichment on four Great Basin grasses. Functional Ecology 1:139–143.

Smith, Z. G. Jr. 1987. California hardwoods: A professional challenge to the resource community. Pp. 1–4 in Plumb and Pillsbury 1987.

Solbrig, O. T. 1996. The diversity of the savanna ecosystem. Pp. 1–27 in Solbrig et al. 1996.

Solbrig, O. T., Medina, E., and Silva, J. F. (eds.). 1996. Biodiversity and Savanna Ecosystem Processes: A Global Perspective. Springer-Verlag, New York.

Sowell, J. B. 1985. A predictive model relating North American plant formations and climate. Vegetatio 60:103–111.

Springfield, H. W. 1976. Characteristics and Management of Southwestern Pinyon-Juniper Ranges: The Status of Our Knowledge. USDA Forest Service Rocky Mountain Research Station Research Paper RM-160. Fort Collins, Colo.

Standiford, R. B. (tech. coord.). 1991. Proceedings of the Symposium on Oak Woodlands and Hardwood Rangeland Management. USDA Forest Service General Technical Report PSW-126. Berkeley, Calif.

Stanley, T. R. Jr. 1995. Ecosystem management and the arrogance of humanism. Conservation Biology 9:255–262.

Steele, R. 1987. Ecological relationships of ponderosa pine. Pp. 71–76 in Baumgartner and Lotan 1987.

Steinauer, E. M., and Bragg, T. B. 1987. Ponderosa pine *(Pinus ponderosa)* invasion of Nebraska sandhills prairie. American Midland Naturalist 118:358–365.

Stephenson, N. L. 1990. Climatic control of vegetation distribution: The role of the water balance. American Naturalist 135:649–670.

Steuter, A. A., and Britton, C. M. 1983. Fire-induced mortality of redberry juniper (*Juniperus pinchotii* Sudw.). Journal of Range Management 36: 343–345.

Steuter, A. A., and McPherson, G. R. 1995. Fire as a physical stress. Pp. 550–579 in Bedunah and Sosebee 1995.

Steuter, A. A., and Wright, H. A. 1983. Spring burning to manage redberry juniper rangelands—Texas Rolling Plains. Rangelands 5:249–251.

Steuter, A. A., Jasch, B., Ihnen, J., and Tieszen, L. L. 1990. Woodland/grassland boundary changes in the middle Niobrara Valley of Nebraska identified by $\delta^{13}C$ values of soil organic matter. American Midland Naturalist 124:301–308.

Stewart, A. W., and Hurst, G. A. 1987. Vegetation in the longleaf-slash pine forest, Biloxi District, Desoto National Forest, Mississippi. Pp. 149–155 in Pearson et al. (comps.) 1987.

Stewart, O. C. 1951. Burning and natural vegetation in the United States. Geographical Review 41:317–320.

Stoddard, H. L. Sr. 1931. The Bobwhite Quail: Its Habits, Preservation, and Increase. Scribner's, New York.

Stokes, M. A., and Dieterich, J. H. (tech. coords.). 1980. Proceedings of the Fire History Workshop, October 20–28, 1980, Tucson, Arizona. USDA Forest Service Rocky Mountain Research Station General Technical Report RM-81. Fort Collins, Colo.

Stone, E. L. Jr., and Stone, M. H. 1954. Root collar sprouts in pine. Journal of Forestry 52:487–491.

Strain, B. R., and Cure, J. D. (eds.). 1985. Direct Effects of Increasing Carbon Dioxide on Vegetation. U.S. Department of Energy, Office of Basic Research, Carbon Dioxide Research Division, Washington, D.C.

Streng, D. R., and Harcombe, P. A. 1982. Why don't East Texas savannas grow up to forest? American Midland Naturalist 108:278–294.

Streng, D. S., Glitzenstein, J. S., and Platt, W. J. 1993. Evaluating effects of season of burn in longleaf pine forests: A critical review and some results from an ongoing long-term study. Proceedings of the Tall Timbers Fire Ecology Conference 18:227–263.

Stuth, J. W. 1991. Foraging behavior. Pp. 65–83 in Heitschmidt and Stuth 1991.

Sumrall, L. B., Roundy, B. A., Cox, J. R., and Winkel, V. K. 1991. Influence of canopy removal by burning or clipping on emergence of *Eragrostis lehmanniana* seedlings. International Journal of Wildland Fire 1:35–40.

Svejcar, T. J., and Brown, J. R. 1991. Failures in the assumptions of the condition and trend concept for management of natural ecosystems. Rangelands 13:165–167.

Swetnam, T. W. 1990. Fire history and climate in the southwestern United States. Pp. 6–17 in Krammes 1990.

Swetnam, T. W., and Baisan, C. H. 1995. Historical fire occurrence in remote mountains of southwestern New Mexico and northern Mexico. Pp. 153–156 in Brown et al. 1995.

Swetnam, T. W., and Baisan, C. H. 1996. Historical fire regime patterns in the southwestern United States since A.D. 1700. Pp. 11–32 in Allen 1996.

Swetnam, T. W., and Betancourt, J. L. 1990. Fire–Southern Oscillation relations in the southwestern United States. Science 262:885–889.

Tajchman, S. J., Keys, R. N., and Kosuri, S. R. 1991. Comparison of pH, sulfate, and nitrate in throughfall and stemflow in yellow-poplar and oak stands in north-central West Virginia. Forest Ecology and Management 40:137–144.

Tanner, G. W. 1987. Soils and vegetation of the longleaf/slash pine forest type, Apalachicola National Forest, Florida. Pp. 186–200 in Pearson et al. (comps.) 1987.

Tausch, R. J., Wigand, P. E., and Burkhardt, J. W. 1993. Viewpoint. Plant community thresholds, multiple steady states, and multiple successional pathways: Legacy of the Quaternary? Journal of Range Management 46:439–447.

Tester, J. R. 1989. Effects of fire frequency on oak savanna in east-central Minnesota. Bulletin of the Torrey Botanical Club 116:134–144.

Thurow, T. L., Blackburn, W. H., Warren, S. D., and Taylor, C. A. Jr. 1987. Rainfall interception by midgrass, shortgrass, and live oak mottes. Journal of Range Management 40:455–460.

Tiedemann, A. R., and Klemmedson, J. O. 1973. Effects of mesquite on physical and chemical properties of the soil. Journal of Range Management 26:27–29.

Tiedemann, A. R., and Klemmedson, J. O. 1977. Effects of mesquite trees on vegetation and soils in the desert grassland. Journal of Range Management 30:361–367.

Tiedemann, A. R., and Klemmedson, J. O. 1986. Long-term effects of mesquite removal on soil characteristics: I. Nutrients and bulk density. Soil Science Society of America Journal 50:472–475.

Tieszen, L. L., and Archer, S. 1990. Isotopic assessment of vegetation changes in grassland and woodland systems. Pp. 293–321 in Osmond et al. 1990.

Tilman, D. 1990. Constraints and tradeoffs: Toward a predictive theory of competition and succession. Oikos 58:3–15.

Tissue, D. T., and Oechel, W. C. 1987. Response of *Eriophorum vaginatum* to elevated CO_2 and temperature in the Alaskan tussock tundra. Ecology 68:401–410.

Tothill, J. C., and Mott, J. C. (eds.). 1985. Ecology and Management of the World's Savannas. Australian Academy of Science, Canberra.

Trabalka, J. R. (ed.). 1986. Atmospheric Carbon Dioxide and the Global

Carbon Cycle. U.S. Department of Energy, Office of Basic Services, Washington, D.C.
Trabalka, J. R., Edmonds, J. A., Reilly, J., Gardner, R. H., and Voorhees, L. D. 1986. Human alterations of the global carbon cycle and the projected future. Pp. 247–302 in Trabalka 1986.
Turner, M. G., and Ruscher, C. L. 1988. Changes in landscape patterns in Georgia, USA. Landscape Ecology 1:241–251.
Turner, R. M. 1990. Long-term vegetation change at a fully protected Sonoran Desert site. Ecology 71:464–477.
Tweney, R. D., Doherty, M. E., and Mynatt, C. R. (eds.). 1981. On Scientific Thinking. Columbia University Press, New York.
Underwood, A. J. 1995. Ecological research and (and research into) environmental management. Ecological Applications 5:232–247.
Vaitkus, M. R., and Eddleman, L. E. 1991. Tree size and understory phytomass production in a western juniper woodland. Great Basin Naturalist 51:236–243.
Vallentine, J. F. 1989. Range Development and Improvements. 3rd ed. Academic Press, San Diego.
Vallentine, J. F. 1990. Grazing Management. Academic Press, San Diego.
Van Auken, O. W., and Bush, J. K. 1987. Influence of plant density on the growth of *Prosopis glandulosa* var. *glandulosa* and *Büchloe dactyloides.* Bulletin of the Torrey Botanical Club 114:393–401.
Van Auken, O. W., and Bush, J. K. 1988. Competition between *Schizachyrium scoparium* and *Prosopis glandulosa.* American Journal of Botany 75:782–789.
Van Auken, O. W., and Bush, J. K. 1989. *Prosopis glandulosa* growth: Influence of nutrients and simulated grazing of *Bouteloua curtipendula.* Ecology 70:512–516.
Van Auken, O. W., and Bush, J. K. 1990. Importance of grass density and time of planting on *Prosopis glandulosa* seedling growth. Southwestern Naturalist 35:411–415.
Vander Wall, S. B., and Balda, R. P. 1977. Coadaptations of the Clark's nutcracker and the piñon pine efficient seed harvest and dispersal. Ecological Monographs 47:89–111.
Vander Wall, S. B., and Balda, R. P. 1981. Ecology and evolution of the food-storage behavior in conifer seed-caching Corvids. Zeitschrift fur Tierpsychologie 56:217–242.
Van Devender, T. W. 1995. Desert grassland history: Changing climates, evolution, biogeography, and community dynamics. Pp. 68–99 in McClaran and Van Devender 1995.
Van Hooser, D. D., and Keegan, C. E. III. 1987. Distribution and volumes of ponderosa pine forests. Pp. 1–6 in Baumgartner and Lotan 1987.
Van Hooser, D. D., O'Brien, R. A., and Collins, D. C. 1993. New Mexico's

Forest Resources. USDA Forest Service Intermountain Research Station Resource Bulletin INT-79. Ogden, Utah.

Vankat, J. L. 1979. The Natural Vegetation of North America. Wiley, New York.

Vavra, M., Laycock, W., and Pieper, R. (eds.). 1994. Ecological Implications of Livestock Herbivory in the West. Society for Range Management, Denver.

Veblen, T. T., and Lorenz, D. C. 1991. The Colorado Front Range: A Century of Ecological Change. University of Utah Press, Salt Lake City.

Veno, P. A. 1976. Successional relationships of five Florida plant communities. Ecology 57:498–508.

Verrall, A. F. 1936. The dissemination of *Septoria acicola* and the effect of grass fires on it in pine needles. Phytopathology 26:1021–1024.

Villanueva-Díaz, J., and McPherson, G. R. 1995. Forest stand structure in mountains of Sonora, Mexico, and New Mexico, USA. Pp. 416–423 in DeBano et al. 1995.

Virginia, R. A., and Jarrell, W. M. 1983. Soil properties in a mesquite-dominated Sonoran Desert ecosystem. Soil Science Society of America Journal 47:138–144.

Wade, D. D., and Johansen, R. W. 1986. Effects of Fire on Southern Pine: Observations and Recommendations. USDA Forest Service Southeastern Research Station General Technical Report SE-42. Atlanta, Ga.

Wahlenberg, W. G. 1946. Longleaf Pine: Its Use, Ecology, Regeneration Protection, Growth, and Management. Charles Lathrop Pack Forestry Foundation, Washington, D.C.

Wakeley, P. C. 1970. Thirty-year effects of uncontrolled brown spot on planted longleaf pine. Forest Science 16:197–202.

Walker, B. H. (ed.). 1987. Determinants of Tropical Savannas. ICSU Press, Oxford, U.K.

Walker, B. H., and Noy-Meir, I. 1982. Aspects of the stability and resilience of savanna ecosystems. Pp. 556–590 in Huntley and Walker 1982.

Walker, B. H., Ludwig, D., Holling, C. S., and Peterman, R. M. 1981. Stability of semiarid savanna grazing systems. Journal of Ecology 69: 473–498.

Walker, B. H., Robertson, J. A., and Penridge, L. K. 1986. Herbage response to tree thinning in a *Eucalyptus crebra* woodland. Australian Journal of Ecology 11:135–140.

Walter, H. 1954. Die verbuschung, eine erscheinung der subtropischen savannengebiete, und ihre ökologischen urscachen. Vegetatio 5/6: 6–10.

Walter, H. 1979. Vegetation of the Earth and Ecological Systems of the Geo-Biosphere. Springer-Verlag, New York.

Ware, S., Frost, C., and Doerr, P. D. 1993. Southern mixed hardwood forest: The former longleaf pine forest. Pp. 447–493 in Martin et al. 1993.

Warren, A., Holechek, J., and Cardenas, M. 1996. Honey mesquite influences on Chihuahuan desert vegetation. Journal of Range Management 49:46–52.

Warren, S. D., Nevil, M. B., Blackburn, W. H., and Garza, N. E. 1986. Soil response to trampling under intensive rotation grazing. Soil Science Society of America Journal 50:1336–1341.

Watts, W. A. 1969. A pollen diagram from Mud Lake, Marion County, north-central Florida. Geological Society of America Bulletin 80: 631–642.

Watts, W. A. 1980. The late Quaternary vegetation history of the southeastern United States. Annual Review of Ecology and Systematics 11: 387–409.

Weaver, H. 1951. Fire as an ecological factor in the southwestern ponderosa pine forests. Journal of Forestry 49:93–98.

Weaver, J. E. 1968. Prairie Plants and Their Environment. University of Nebraska Press, Lincoln.

Webb, T. III, and Bartlein, P. J. 1992. Global changes during the last 3 million years: Climatic controls and biotic responses. Annual Review of Ecology and Systematics 23:141–173.

Webber, H. J. 1935. The Florida scrub, a fire fighting association. American Journal of Botany 22:344–361.

Weiner, J. 1995. On the practice of ecology. Journal of Ecology 83:153–158.

Weiss, C. (ed.). 1972. Evaluating Action Programs. Allyn and Bacon, Boston.

Weldon, D. 1986. Exceptional physical properties of Texas mesquite wood. Forest Ecology and Management 16:149–153.

Welker, J. M., and Menke, J. W. 1990. The influence of simulated browsing on tissue water relations, growth, and survival of *Quercus douglasii* (Hook and Arn.) seedlings under slow and rapid rates of soil drought. Functional Ecology 4:807–817.

Wells, B. W. 1928. Plant communities of the coastal plain of North Carolina and their successional relations. Ecology 9:230–242.

Wells, B. W., and Shunk, I. V. 1931. The vegetation and habitat factors of coarser sands of the North Carolina coastal plain: An ecological study. Ecological Monographs 1:465–520.

Wells, P. V. 1970. Postglacial vegetational history of the Great Plains. Science 167:1574–1582.

Wells, P. V. 1983. Paleobiogeography of montane islands in the Great Basin since the last glaciopluvial. Ecological Monographs 53:341–382.

Weltzin, J. K., and McPherson, G. R. 1995. Potential effects of climate change on lower treelines in the southwestern United States. Pp. 180–193 in DeBano et al. 1995.

Weltzin, J. K., Archer, S., and Heitschmidt, R. K. 1997. Small-mammal regulation of vegetation structure in a temperate savanna. Ecology 78:751–763.

Werger, M. J. A., van der Aart, P. J. M., During, H. J., and Verhoeven, J. T. A. (eds.). 1988. Plant Form and Vegetation Structure. SPB Academic Publishing, The Hague, Netherlands.

Werner, P. A., Walker, B. H., and Stott, P. A. 1990. Introduction. Journal of Biogeography 17:343-344.

West, N. E. 1984. Successional patterns and productivity potentials of pinyon-juniper ecosystems. Pp. 1301–1332 in National Research Council/National Academy of Sciences, Developing Strategies for Rangeland Management. Westview Press, Boulder, Colo.

West, N. E. 1988. Intermountain deserts, shrub steppes, and woodlands. Pp. 209–230 in Barbour and Billings 1988.

West, N. E., and Van Pelt, N. S. 1987. Successional patterns in pinyon-juniper woodlands. Pp. 43–52 in Everett 1987.

West, N. E., Tausch, R. J., Rea, K. H., and Southard, A. R. 1978. Soils associated with pinyon-juniper woodlands of the Great Basin. Pp. 68–88 in Youngberg 1978.

Westoby, M., Walker, B., and Noy-Meir, I. 1989. Opportunistic management for rangelands not at equilibrium. Journal of Range Management 42:266–274.

Whetten, N. L. 1948. Rural Mexico. University of Chicago Press, Chicago.

Whicker, A. D., and Detling, J. K. 1988. Ecological consequences of prairie dog disturbances. Bioscience 38:778–785.

White, A. S. 1985. Presettlement regeneration patterns in a southwestern ponderosa pine stand. Ecology 66:589–594.

White, K. L. 1966. Structure and composition of foothill woodland in central coastal California. Ecology 47:229–237.

White, S. S. 1948. The vegetation and flora of the Rio de Bavispe in northeastern Sonora, Mexico. Lloydia 11:229–302.

White, T. C. R. 1993. The Inadequate Environment: Nitrogen and the Abundance of Animals. Springer-Verlag, Berlin.

Whitham, T. G., Maschinski, J., Larson, K. C., and Paige, K. N. 1991. Plant responses to herbivory: The continuum from negative to positive and underlying physiological mechanisms. Pp. 227–256 in Price et al. 1991.

Wicklum, D., and Davies, R. W. 1995. Ecosystem health and integrity? Canadian Journal of Botany 73:997–1000.

Wigley, T. M. L. 1985. Impact of extreme events. Nature 316:106–107.

Williams, J. D. 1991. The U.S. population, 1970 to 2010: Size, geographic distribution, and age structure. Pp. 25–50 in Demographic Change and the Economy of the Nineties. Congressional Research Service, Library of Congress. GPO, Washington, D.C.

Williams, W. E., Garbutt, K., Bazzaz, F. A., and Vitousek, P. M. 1986. The response of plants to elevated CO_2: IV. Two deciduous-forest tree communities. Oecologia 69:454–459.

Willson, M. F. 1981. Commentary: Ecology and science. Bulletin of the Ecological Society of America 62:4–12.

Wink, R. L., and Wright, H. A. 1973. Effects of fire on an Ashe juniper community. Journal of Range Management 26:326–329.

Witty, J. E., and Knox, E. G. 1964. Grass opal in some chestnut and forested soils in north-central Oregon. Soil Science Society of America Proceedings 28:685–688.

Wolters, G. L. 1981. Timber thinning and prescribed burning as methods to increase herbage on grazed and protected longleaf pine ranges. Journal of Range Management 34:494–497.

Wondzell, S., and Ludwig, J. A. 1995. Community dynamics of desert grasslands: Influences of climate, landforms, and soils. Journal of Vegetation Science 6:377–390.

Wong, S. -C. 1979. Elevated atmospheric partial pressure of CO_2 and plant growth: I. Interactions of nitrogen nutrition and photosynthetic capacity in C_3 and C_4 plants. Oecologia 44:68–74.

Wray, S. B., and Strain, B. R. 1987. Competition in old-field perennials under CO_2 enrichment. Ecology 68:1116–1120.

Wright, H. A. 1974. Range burning. Journal of Range Management 27:5–11.

Wright, H. A., and Bailey, A. W. 1982. Fire Ecology: United States and Southern Canada. Wiley, New York.

Wright, H. A., and Klemmedson, J. O. 1965. Effects of fire on bunchgrasses of the sagebrush-grass region in southern Idaho. Ecology 46: 680–688.

Wright, H. A., Bunting, S. C., and Neuenschwander, L. F. 1976. Effect of fire on honey mesquite. Journal of Range Management 29:467–471.

Wright, H. A., Neuenschwander, J. F., and Britton, C. M. 1979. The Role and Use of Fire in Sagebrush-Grass and Pinyon-Juniper Plant Communities: A State-of-the-Art Review. USDA Forest Service Intermountain Research Station General Technical Report INT-58. Ogden, Utah.

Wright, R. A. 1982. Aspects of desertification in *Prosopis* dunelands of southern New Mexico, U.S.A. Journal of Arid Environments 5:277–284.

Wright, R. G., and Van Dyne, R. G. 1981. Population age structure and its relationship to the maintenance of the semidesert grassland undergoing invasion by mesquite. Southwestern Naturalist 26:13–22.

Young, J. A., and Evans, R. A. 1981. Demography and fire history of a western juniper stand. Journal of Range Management 34:501–506.

Young, J. A., and Evans, R. A. 1987. Stem flow on western juniper *(Juniperus occidentalis)* trees. Pp. 373–381 in Everett 1987.

Young, J. A., and Young, C. G. 1992. Seeds of Woody Plants in North America. Dioscorides Press, Portland, Ore.

Young, J. A., Evans, R. A., and Easi, D. A. 1984. Stem flow on western juniper *(Juniperus occidentalis)* trees. Weed Science 32:320–327.

Young, J. A., Evans, R. A., and Weaver, R. A. 1976. Estimating potential downy brome competition after wildfires. Journal of Range Management 29:322–325.

Young, M. D., and Solbrig, O. T. (eds.). 1993. The World's Savannas. Parthenon Publishing Group, Pearl River, New York.

Youngberg, C. T. (ed.). 1978. Proceedings of the Fifth North American Forest Soils Conference. Colorado State University, Fort Collins.

Index

About the Author

Guy McPherson is an associate professor in the School of Renewable Natural Resources at the University of Arizona. He grew up in northern Idaho and received an undergraduate degree in forestry from the University of Idaho. He completed his formal education at Texas Tech University, where he received an M.S. and a Ph.D. in range science. He moved to Tucson after conducting postdoctoral research at the University of Georgia and teaching at Texas A & M University.

Results of McPherson's field research on savannas, woodlands, and grasslands have been published in over 30 journal articles and book chapters. McPherson has authored or coauthored articles in most major journals devoted to the ecology and management of natural resources, including *Oecologia, Journal of Applied Ecology, Journal of Vegetation Science, Journal of Range Management,* and *Forest Ecology and Management.*

McPherson's recent and ongoing research focuses on mechanisms of vegetation change and plant-animal interactions in southwestern oak savannas and semidesert grasslands. This research includes explicit assessment of the effects of disturbance and climate change on the structure and function of ecosystems. Through this research McPherson has addressed fundamental issues in ecology, aiming at practical applications; he has thus incorporated the major tools available to resource managers to alter species composition, including prescribed fire, livestock grazing, and mechanical and chemical treatments.